Contents

Summary	vi
Résumé	vii
Acknowledgements	viii

Part 1 ENVIRONMENTAL STATEMENT	1
Introduction	3
Tsetse fly eradication in Zimbabwe	3
Study design	4
Box 1 Ground-spraying operations with DDT	7
Box 2 Scope of the Assessment	8
Box 3 The Siabuwa study area	8
Ecological Assessment	14
Terrestrial environment	14
Aquatic environment	19
General conclusions	22
Risk to Human Health	22
Economic Assessment	23
Mitigation Measures	23
Alternative insecticide	23
Alternative control techniques	23
Conclusions	24
Project Publications	25
References	26
Part 2 SCIENTIFIC REPORT	29
Effects on Wildlife: Terrestrial Studies	31
1 **Deposition and Dissipation**	33
I F Grant	
Spray drift	33
Residue levels on bark and in soil	33
Dissipation of residues from bark and soil	35
References	35

2	**Critical Soil Processes**	39

I F Grant

Litter degradation — 39
Soil respiration — 43
Nitrification — 43
Discussion — 45
Conclusions — 46
References — 47

3	**Terrestrial Invertebrates**	49

C C D Tingle

Introduction — 49
Study rationale — 50
Invertebrates of mopane woodland — 52
The effects of DDT — 58
Invertebrates in the food chain — 67
Discussion — 69
References — 69

4	**Lizards**	73

M R K Lambert

Introduction — 73
Lizards in woodland — 73
Lizards in rock outcrops — 76
Discussion and conclusions — 77
References — 79

5	**Woodland Birds**	81

R J Douthwaite

Introduction — 81
Effects on songbirds — 82
Effects on birds of prey — 97
References — 100

6	**Nocturnal Animals**	103

A N McWilliam

Bats — 103
DDT residues in bats — 112
Population effects — 115
Relative abundance — 117
Bat numbers — 121
Bat activity — 125
Other nocturnal insectivores and carnivores — 127
Conclusions — 129
References — 131

Effects on Wildlife: Aquatic Studies — 135

7 Fish — 137
B McCarton and W Mhlanga
- Introduction — 137
- The fish fauna — 139
- The fishery — 140
- Effects of DDT on fish — 140
- Contamination route and dynamics — 141
- Effects of DDT on growth and mortality — 142
- Growth and mortality in the Songu river — 144
- Growth and mortality in Lake Kariba — 151
- Conclusions — 164
- References — 165

8 Waterbirds — 169
R J Douthwaite
- The fish eagle — 169
- The reed cormorant — 174
- References — 175

Other Effects — 179

9 Risk to Human Health — 181
P Goll and B McCarton
- Effects in Man — 181
- Regulatory limits — 182
- Exposure from foodstuffs — 182
- Conclusions — 184
- References — 184

10 Economic Assessment — 187
P Abelson
- Food resources — 187
- Human Health — 187
- Recreational resources — 188
- Ecosystem impacts — 189
- Existence values — 189
- Conclusion — 189
- References — 189

11 The International Dimension — 191
P Goll
- Global use of DDT today — 191
- References — 192

12 Deltamethrin: a safer substitute for ground-spraying? — 193
R J Douthwaite
- Terrestrial effects — 193
- Aquatic effects — 195
- References — 195

SUMMARY

The environmental impact of DDT used in ground-spraying operations to eradicate tsetse flies from north western Zimbabwe was assessed in field studies between 1987–1991.

Fauna monitored included populations of bats, birds, lizards, fish and insects; microbial processes contributing to soil fertility were also checked. Despite the relatively low application rate, and rapid dissipation of residues, adverse effects on a landscape scale were found in populations of four bird and one lizard species. The comparative scarcity of several bird and terrestrial invertebrate species in sprayed areas may also have been due to DDT. Residue concentrations in at least 5 bat species posed a significant risk to survival during drought. No significant effects were detected in fish or on soil processes. Effects on the majority of scarcer species are unknown.

The effects are reversible, probably within 10–20 years, and are less serious than those caused by habitat loss due to human settlement and elephant damage.

In economic terms, the environmental damage cost of using DDT for tsetse fly control was very low. DDT residue burdens in humans were not monitored in this study but the domestic use of DDT for mosquito control is probably a far more important source of exposure in man.

The adverse effects of ground-spraying with DDT can be mitigated by alternative control techniques or by substituting deltamethrin, a less persistent insecticide, for DDT. However, if substitution increases costs significantly, wildlife conservation would benefit more from the retention of DDT and investment of savings in projects to manage wildlife habitat than from substitution.

Résumé

Des études ont été effectuées sur le terrain de 1987 à 1991 pour évaluer l'impact sur l'environnement du DDT pulvérisé au cours d'opérations au sol pour éradiquer les mouches tsé-tsé des régions nord-ouest du Zimbabwé.

La faune ainsi surveillée comprenait des populations de chauve-souris, d'oiseaux, de lézards, de poissons et d'insectes; les processus microbiens qui contribuent à la fertilité du sol ont aussi été vérifiés. En dépit du taux d'épandage relativement faible et de la dispersion rapide des résidus, on trouva des effets nocifs à une échelle globale dans les populations de quatre espèces d'oiseaux et d'une de lézards. La rareté comparative de plusieurs espèces d'oiseaux et d'espèces d'invertébrés terrestres dans les terrains pulvérisés pouvait aussi résulter du DDT. Les concentrations de résidus dans au moins cinq espèces de chauve-souris faisaient planer un risque significatif sur la survie pendant la sécheresse. Aucun effet significatif n'a été dépisté sur les poissons, ni sur les processus édaphiques. Les effets sur la majorité des espèces les plus rares sont inconnus.

Les effets sont réversibles, probablement dans un délai de 10 à 20 ans, et moins graves que ceux qui résultent des pertes d'habitat provoquées par l'implantation humaine et les dégâts des éléphants.

En termes économiques, les dégâts causés à l'environnement par l'usage du DDT contre la tsé-tsé étaient très faibles. Les résidus de DDT chez l'humain n'ont pas été surveillés au cours de cette étude mais l'usage domestique du DDT pour la lutte contre les moustiques est probablement une source bien plus importante d'exposition à ce produit chez l'humain.

Les effets nocifs de la pulvérisation de DDT au sol peuvent être mitigés par d'autres techniques de lutte ou par le remplacement du DDT par la deltaméthrine, insecticide moins rémanent. Toutefois, si ce remplacement accroît les coûts à un point significatif, la conservation de la faune bénéficiera davantage du maintien de l'usage de DDT et de l'investissement des économies dans des projets de gestion de la faune et de la flore sauvages que d'un tel remplacement.

ACKNOWLEDGEMENTS

It is a pleasure to thank the many people who assisted this project and especially Mr B. Hursey, former Director of Tsetse Control Branch, Dr G. Vale and Mr V. Chadenga, his successors, and Mr W. Shereni, Chief Glossinologist. Their consistent support was matched by that of the late Mr M. Watson and Mr A. Tainsh, Senior Natural Resources Advisers of the Overseas Development Administration.

Fieldwork depended heavily upon others and notably Dr H.P.Q. Crick, Mr F. Cotterell, Mr J. Diza, Mrs B. Douthwaite, Miss M.R. Douthwaite, Mr A. Gusha, Mr K. Hustler, Mr Z. Kapfumo, Mr P. Karasa, Mr L. Kubanga, Mr H. Mafara, Mr K. Mujinga, Mr. W. Muzamba, Mr H. Ncube, Miss A. Popovich, Mr M. Sibanda, Mr T. Sithole, Miss C. Smith, Miss C. Towner, Miss S. Wales-Smith, Mrs E. Wood and the Zimbabwe Falconers' Club.

Thanks are also due to Mr R. Allsopp, Dr G. Armstrong, Miss E. Barnett, Mr J. Barrett, Dr R. Beales, Dr B. Blake, Mr B. Bolton, Dr R. Booth, Dr D.G. Broadley, Dr R. Cheke, Dr S. Chimbuya, Miss A. Cobb, Dr A. Coneybeare, Mr J. Cox, Dr M. Cox, Dr C. Craig, Dr B.R. Critchley, Dr J.M. Dangerfield, Dr R. Davies, Mr M. Day, Mr J. Deeming, Dr R.H. Disney, Miss J. Ellis, Prof. M.B. Fenton, Mr C. Gay, Dr D. Gibson, Dr K. Harris, Dr C. Hodgson, Dr R. Jocque, Mr A. Kumirai, Miss S. Lauer, Mrs J. Laurence, Mr D. F. Lovemore, Prof. J.P. Loveridge, Mr D. MacFarlane, Dr D.S. Madge, Dr R. Madge, Dr J. Magor, Dr R. Martin, Mr A. Masterson, Mr P. Matthiessen, Dr S. Mazur, Dr J. Minshull, Mr B. Muchenje, Dr P. Mundy, Mr J. Murphy, Dr J. Noyse, Mr J. Perfect, Dr R.J. Phelps, Prof. P.A. Racey, Mr F. Robinson, Dr A. Russell-Smith, Mr A. Serle, Mr B. Schoeman, Dr S. Taiti, Dr J. Tarbit, Dr R. Taylor, Dr S.R. Telford, Mr L. Toet, Dr B. Turner, Dr J.A. Wallwork, Dr I. White and Mr T. Wijers who helped in a variety of ways and, finally, to Mrs K. Whitwell and Mr P. Birkett, for editing and presenting our disparate reports as a coherent whole.

Part 1
Environmental Statement

INTRODUCTION

Tsetse Fly Eradication in Zimbabwe

Zimbabwe's population numbers some nine million people and has one of the highest growth rates in Africa. Land hunger and the inequity of land distribution have been important political issues both before and after independence and resettlement and rural agricultural development remain important government priorities.

Traditional agro-pastoralism has been restricted by tsetse flies, *Glossina* spp., which spread the debilitating, and sometimes fatal parasitic disease trypanosomiasis to man and his livestock (Plate 1). Control measures, begun in 1933, have pushed back the limits of infestation to the Zambezi valley and Zimbabwe's borders with Zambia and Mocambique, but about a third of the country is at risk from re-invasion (Figure 1).

Eradication has been achieved by a variety of means involving game elimination (1933–60), which deprived the flies of food, and insecticides (1958–present) (Plates 2–4). Ground spraying with DDT was introduced in 1967 and proved to be the most successful technique, resulting in the eradication of tsetse flies from over 50 000 km^2 of land (Figure 2, Box 1). However, with growing public opposition to the use of DDT, and the recent development of alternative control techniques, the government of Zimbabwe commissioned the present assessment as part of a wider review of the options for change.

DDT was registered for agricultural use in Southern Rhodesia (now Zimbabwe) in 1946 and its use extended to tsetse fly control in 1967 and mosquito control in 1972. In the early 1970s consumption probably reached about 800 tonnes per annum but following restrictions elsewhere, it was withdrawn from domestic and garden use in 1973 and 'banned' in 1985 except for research work related to tsetse fly and mosquito control[3,4]. Use for 'research' has declined each year from 443 tonnes in 1986 to zero in 1991 (Figure 3).

Studies between 1972 and 1983 showed that contamination of Zimbabwean wildlife with DDT residues was widespread[5–10]. Potentially harmful concentrations of DDE were found in eggs of several predatory bird species and in 1981, Provincial Game Warden, W.R.Thomson, predicted extinction of the Fish Eagle, Peregrine and Lanner Falcons, Black Sparrowhawk and other raptors, herons and cormorants within 10–15 years unless the use of DDT was curtailed[11]. He also revealed that DDT residue concentrations in the breast milk of some Zimbabwean women were as high as anywhere else in the World[12].

In a preliminary assessment, Matthiessen[13,14] found no accumulation in the physical environment, but residues were widespread and there was evidence of accumulation in some animals. Though none was at immediate risk from acute poisoning, some bats and some birds, particularly birds of prey, were at risk from sub-lethal poisoning, as were fish fry in rivers draining sprayed areas. Tsetse control operations were the main source of residues in the areas he sampled and he recommended that the biological impact of DDT residues on wildlife, particularly birds, bats, fish and insects, should be studied.

Matthiessen's recommendation was accepted and the present project commissioned in 1987 (Box 2). The general objective was "to produce a better understanding of the environmental costs of using DDT, or alternative insecticides, for ground spraying against tsetse and to provide evidence which contributes to a rational evaluation of the role of this technique in integrated campaigns for tsetse fly control."

Study Design

The lack of base-line data on the effects of DDT on wildlife populations in the Tropics and about the animal communities present in the project area were major constraints and necessitated a considerable amount of field-based research, which extended from 1987 to 1991.

DDT can have a variety of effects on wildlife — lethal or sub-lethal, acute or chronic, direct or indirect — acting upon the individual animal, species population, or entire community. The most serious are those affecting species populations or ecosystem processes at a landscape scale. Priority in this assessment was therefore given to studying vertebrate populations, which would integrate effects on a landscape scale, rather than invertebrates, which would be more likely to show site-specific responses to DDT. In addition, anthropomorphic judgements would generally rate effects on vertebrates more serious than similar effects in invertebrates. However, studies were also made on soil fertility and arthropods, to check whether spraying operations led to fundamental changes in productivity with indirect consequences for vertebrate populations.

The assessment team comprised a full-time ornithologist/team leader, a terrestrial ecologist, specializing in bats, and a fish biologist, who were assisted by short-term inputs from a second ornithologist, a herpetologist, entomologist and soil scientist.

The study approach was to compare the relative abundance of species populations in unsprayed and heavily sprayed areas and to monitor changes in relation to on-going operations. Possible reasons for scarcity or decline in sprayed areas were then investigated. The lack of previous ecological work in the area increased the difficulty of this task.

By 1987, spraying operations had covered much of the Zambezi valley and the choice of study areas was very limited. The requirement that unsprayed and heavily sprayed areas should be adjacent, to reduce the confounding effects of environmental variation, and for access to a wide range of sample sites, restricted the choice further. Only one area, north of Siabuwa, was suitable (Figure 4), and most of the work was done here (Box 3). Good luck, rather than design, ensured that spraying operations continued in the area for most of the study period, enabling changes in previously unsprayed areas to be monitored.

The annual spraying operation covers an extensive block of land. Treatment is unreplicated and non-random, and the effects of spraying cannot therefore be distinguished from those of other environmental variables by statistical means. We reduced the confounding effects of environmental variables by choosing similar sites in sprayed and unsprayed areas and we interpreted differences between these areas on the basis of DDT residue concentrations and the associated hazards. The alternative, through randomized sampling of past spray blocks, would have been logistically impossible. Experimental treatments were not an option.

Pesticide residues analysis of bark, soil and silt samples was carried out at an early stage in the project to confirm the contamination status of sites and to monitor residue decay. However, in view of the high cost of residues analysis and the problem of

Environmental Statement

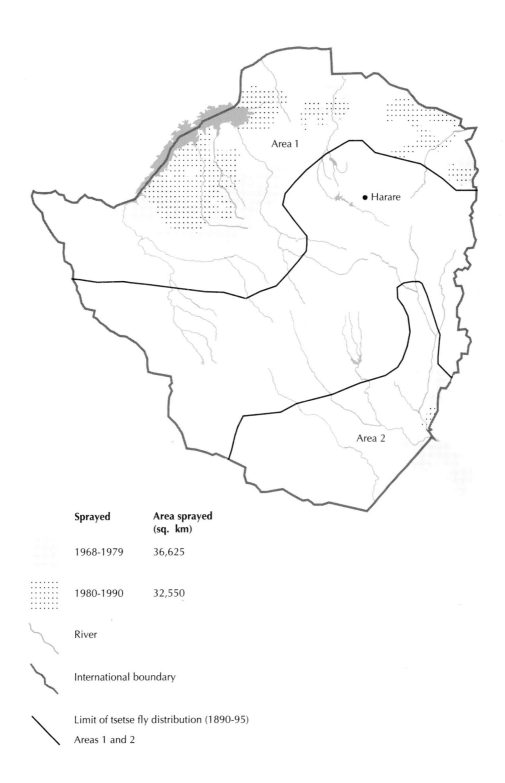

Figure 1 Historical limit of tsetse fly distribution and areas ground-sprayed between 1968–1979 and 1980–1990 in Zimbabwe

DDT in the Tropics

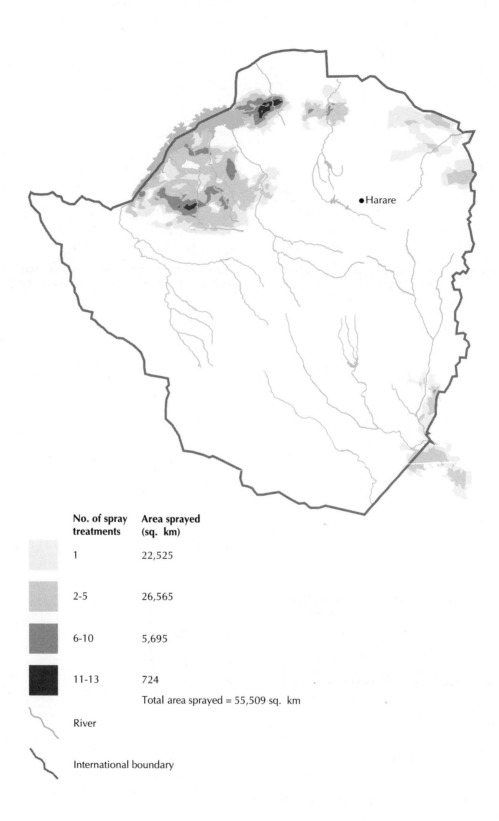

Figure 2 Number of DDT spray treatments between 1968–1990. (The following operations are excluded as the original maps have been lost — 1970: Gatshegatshe (80 km²); 1971: Gatshegatshe (88 km²), Kariba Airport (8 km²), Sabi-Turgwe (200 km²), Mozambique (705 km²); 1973: Mozambique (734 km²); 1978: Hurungwe (306 km²); 1979: Hurungwe (313 km²))

inferring biological impact from residue concentrations in animal tissue, analysis of animal material was generally deferred until fieldwork had shown differences between species populations in sprayed and unsprayed areas.

Box 1 Ground-spraying operations with DDT

A prerequisite for successful ground spraying is adequate access. In uninhabited areas, tracks are bulldozed so that spraying teams never work further than 3 km from the vehicle carrying insecticide.

A second prerequisite is an exact spraying plan. The glossinologist in charge of the operation marks aerial photographs with wax pencil to route spraying teams through essential habitat, through thicket and closed woodland, below escarpments, along drainage lines, field edges, roads, tracks, well-used game and cattle paths, and past cattle kraals, dips and inspection races. The marked photographs are then issued to the leaders of spraying teams to follow.

Spraying is carried out between June and September, in the dry Winter and early Spring. The spraying teams camp close to a reliable source of water. The insecticide spray is mixed and equipment cleaned nearby, with the washings draining to a deep pit not less than 50 m from the river bank. The pit is refilled with soil at the end of the spraying operation.

Reliable transport is essential for the success of the operation (Plate 6). Each spraying team requires a 5-tonne lorry and for every four teams there is an officer-in-charge with a Landrover. A spraying team normally comprises a Field Assistant-in-Charge, an assistant to help him guide the operators, one driver, eight spray operators (who take it in turns to operate the team's four knapsack compression sprayers), eight insecticide carriers and a few odd-job men (Plate 5).

Each spray man covers a band about 15 m wide so that the team's four sprayers have a swath-width capacity of 60 m (Plate 7). The team may be split to cover narrow bands of essential habitat along streams, valleys and tracks. DDT is applied selectively to tsetse resting sites and refuges. Resting sites include large tree trunks and the underside of branches, up to a height of 3–4 m (Plate 8). Special attention is paid to trees on termite mounds, around springs and water holes and those with dark bark. Refuge sites include the underside of fallen logs, rot holes in trees, antbear and warthog holes, rock overhangs and holes in river banks.

Generally about a fifth of the total area falls within a spray swath. The spray contains 4% active ingredient DDT in suspension, made by mixing DDT 75% wettable powder with water. Application rates average about 160–250 g a.i./ha overall, 1 kg/ha in the spray swath, or $0.005 - 1.0$ g/m² on the spray target. In the absence of rain and fire the deposit remains effective against tsetse for up to six months.

Tsetse have been eradicated from over 50 000 km² of land by ground-spraying since 1967 and operations in northern Zimbabwe have moved progressively from the watershed towards the Zambezi river (Figure 1). Most of the area cleared required one or two treatments, but some areas have been treated up to 13 times (Figure 2). An estimated 3837 tonnes of technical DDT (75% w.p.) has been used (J. Barrett personal communication) and annual consumption since 1980 is shown in Figure 3.

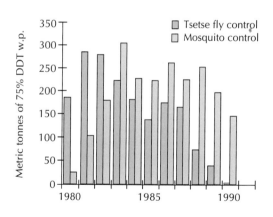

Figure 3 DDT consumption for tsetse fly and mosquito control in Zimbabwe, 1980–1991

Box 2 Scope of the Assessment

The terms of reference were to determine the effects of ground-spraying with DDT and alternative insecticides:

- on organic matter decomposition and mineralization and the consequences for nutrient cycling and soil productivity;
- on terrestrial fauna occupying the same habitat as tsetse, and especially effects on the abundance and diversity of 'keystone' species and groups and their predators, particularly birds and bats;
- on aquatic fauna of seasonal rivers flowing into Lake Kariba, and particularly breeding success in commercially important fish species, and to consider whether any effects are likely to have long-term consequences for the productivity of the lake;
- on reproductive success and eggshell thinning in the Fish Eagle;
- on fish as food for man.

The assessment was thus aimed at determining the effects of insecticides on non-target fauna rather than the broader impact of ground-spraying operations on the environment.

Box 3 The Siabuwa Study Area

Treatment history The northern part of the study area was treated with DDT in 1984 (@ 162 g a.i./ha), 1985 (@ 210 g a.i./ha) and 1986 (@ 247 g a.i./ha) (Figure 5). Tsetse flies had been eradicated to the south, in 1984 and 1985, by aerial spraying with endosulfan[15] and the area was free of DDT except for background contamination. Drift sprays of endosulfan applied at dose rates of 14–25 g a.i./ha can kill small fish[16] and may affect feeding success in piscivorous and insectivorous birds[17,18], but detailed studies in Zimbabwe, Botswana and Somalia found effects are short-lived and ecologically insignificant[19].

Ground-spraying operations with DDT resumed in June 1987 (@ 185 g a.i./ha), encroaching onto land previously unsprayed. Further spraying took place in 1988 (@ 201 g a.i./ha) and 1989 (@ 258 g a.i./ha).

Climate The climate is characterized by a hot, erratic rainy season, between the transitional months of October and April, and a cool dry season, between May and September (Figures 6–7). Rainfall at Siabuwa averaged 664 mm a year over the 40 years 1951–1990, with a coefficient of variation of 38%. Rainfall tends to increase northwards, towards Lake Kariba, but records from May 1988 to May 1991 from two project weather stations indicate no significant variation between the northern and southern ends of the study area.

The project started during a severe drought, following failure of the 1986/87 rains, but over the study period rainfall was close to average, with January and February the wettest months. The rains of 1988/89 were well above average (Table 1).

Temperatures usually peak in October and November before the arrival of the Summer rains, the daily maximum averaging 36°C at this time (Figure 7). They reach their nadir during the dry Winter season in July, when daily minimum temperatures average 13°C.

Hydrology, soils, vegetation and land use The study area lies on the watershed between the Mwenda and Sengwa rivers (Plates 9–12). Much of the area is drained by the Songu river and shorter tributaries of the Sengwa. The Songu rises in the

Table 1 Annual rainfall totals (October–September, mm) affecting the study period

	1986/87	1987/88	1988/89	1989/90	1990/91
Chunga*	510	788	1023	854	556
Siabuwa/Nabusenga Dam (project camp)	360*	627*	839	694	559

* Derived from Zimbabwe Meterological Department Records

Environmental Statement

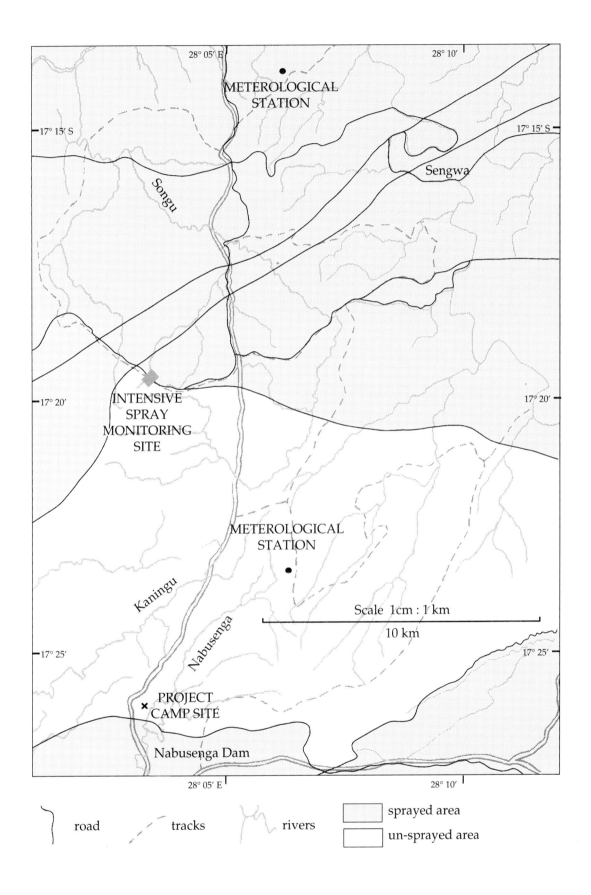

Figure 4 Map of the study area at Siabuwa

DDT in the Tropics

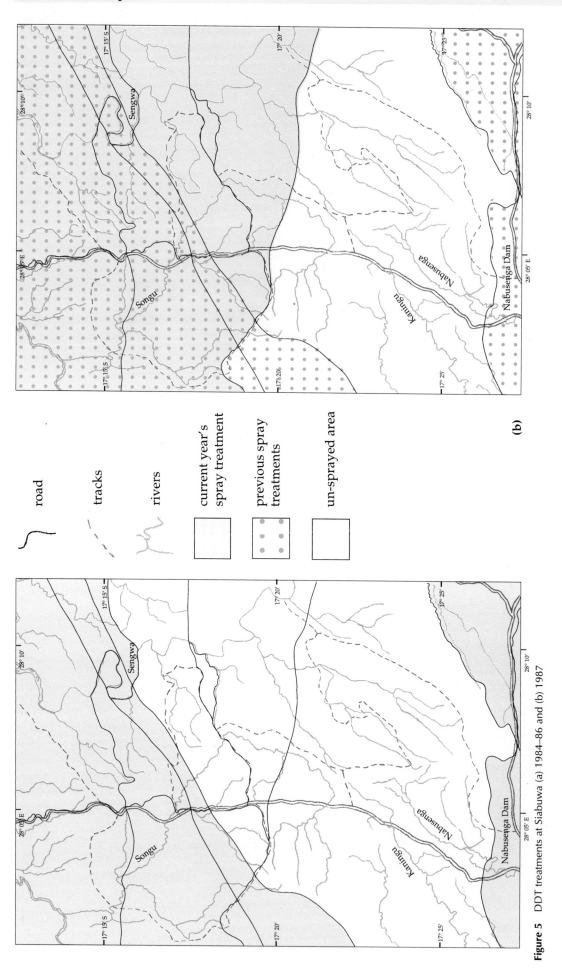

Figure 5 DDT treatments at Siabuwa (a) 1984–86 and (b) 1987

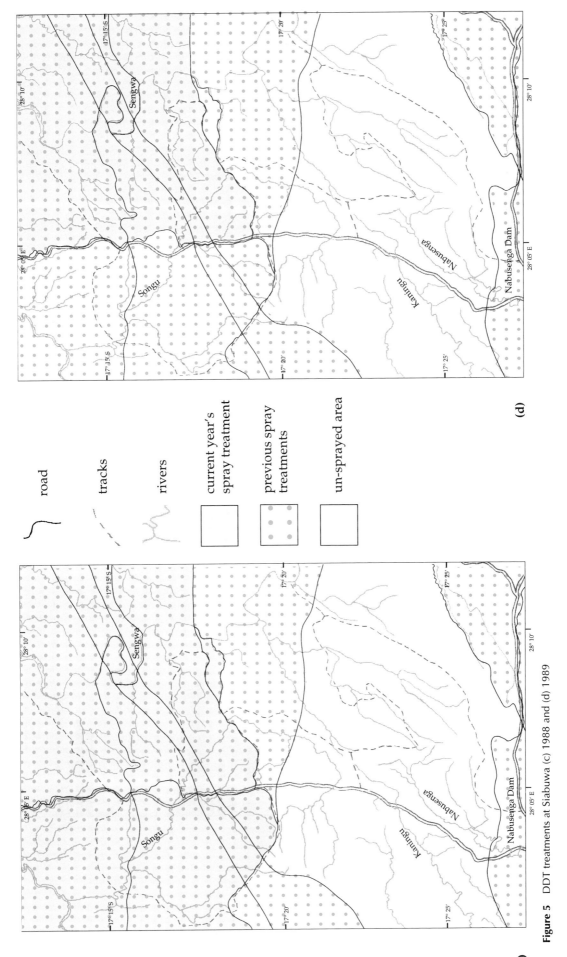

Figure 5 DDT treatments at Siabuwa (c) 1988 and (d) 1989

unsprayed south before entering the sprayed section and then leaving the study area at Siazele Gorge. Its bed is rocky or sandy, scoured by flash floods in the rains. Flow persists intermittently between January and July, depending on the rains, and a few muddy pools survive the dry season in the wetter years.

The vegetation of the area is generally a low, often open, mixed woodland, developed on shallow soils derived from Karroo sandstone. The sandstone varies in lithography, from coarse, flat-bedded grits (as seen on ridges) to fine-grained sands and siltstones (as seen in the more rocky areas with heavier soils), but variation in the vegetation is principally due to degree of dissection and position on slope, realizing its effect through soil depth and texture. The topslopes have deeper sandier soils, while the midslopes are often pebbly or rocky. Bottomslopes consist of heavier clay soils, sometimes stone-covered.

Hill tops are dominated by *Julbernardia globiflora*, *Pseudolachnostylis maprouneifolia*, *Combretum zeyheri* and *Strychnos cocculoides* woodland and a well-developed and tall (75 cm) grass cover, with *Stereochloena cameronii*, *Andropogon* spp., *Heteropogon melanocarpus*, which is often burnt. *Diplorhynchus condylocarpon*, *Baphia massaiensis* and the fire-resistant *Xeromphis obovata*, are common shrubs. Locally, *Pterocarpus angolensis* occurs on sand, and *Burkea africana* and *Baphia massaiensis* in moister areas. Mopane trees, often large, are found on termitaria, or in other areas where the heavier subsoil is exposed. Most species associated with the topslopes become scarcer lower down. Here, mopane is dominant, with *Terminalia stuhlmannii* on rocky areas, and *Terminalia prunoides* on heavier soils. *Diospyros quiloensis* is always associated with mopane. Thickets of *Pteleopsis anisoptera*, *Combretum elaeagonoides*, *Stychnos madagascariensis* and *Hippocratea parvifolia* occur in some areas. On stony areas, *Julbernardia globiflora* is often dominant with *Pteleopsis anisoptera* and *Combretum apiculatum* forming a low, open woodland. On rocky areas *Terminalia stuhlmannii* is common with *Commiphora* spp., *Croton gratissimus* and *Combretum apiculatum*. Ground cover is generally more patchy. Species of *Aristida*, *Sporobolus*, *Eragrostis*, *Digitaria* and *Chloris* are common grasses and *Duosperma crenatum*, *Justicia betonica*, *Bidens schimperi* and *Calostephane divaricata* common herbs.

A few hundred elephants are now present in the area. Past damage to the woodland appears to have been less severe than in surrounding areas but during the study period elephants destroyed large numbers of mature trees in some areas, transforming the woodland.

Villagers were resettled at Siabuwa and Chunga in the 1970s, during the liberation war, and the study area is uninhabited. The old fields and settlements are now marked by thickets of *Combretum elaeagnoides*.

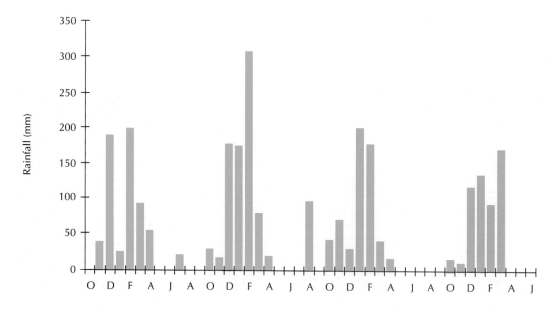

Figure 6 Monthly rainfall totals for Siabuwa, October 1987–June 1991.

Environmental Statement

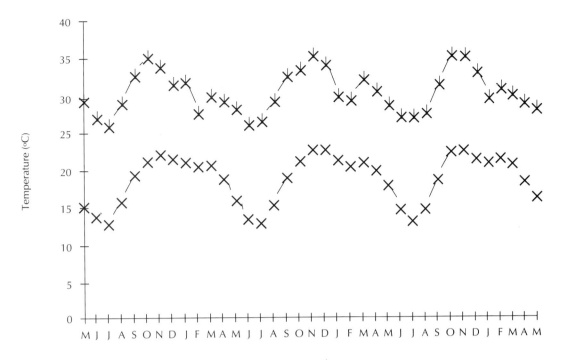

Figure 7 Mean monthly temperatures, May 1988–May 1991. ✻ , Mean daily maximum; ✕ , Mean daily minimum

Figure 8 Meteorological data were logged at two sites in the study area and down-loaded each month to a portable computer for analysis (A.N. McWilliam)

Ecological Assessment

Terrestrial Environment

Distribution and dynamics of DDT

1. Tsetse are normally eradicated with one or two spray treatments, but up to 13 treatments have been made in some areas. One treatment is made annually.

2. The overall application rate amounts to 160–250 g a. i./ha but DDT is concentrated on spray targets clustered within spray swaths. Deposition amounts to 1 kg/ha within the 60 m wide spray swath, or 0.005–1.0 g/m² on the spray target. Average ΣDDT* residues on mopane tree bark from sprayed areas ranged from 0.5–93 µg ΣDDT cm², compared with <0.01–0.03 µg DDT cm² on bark from the unsprayed area. These rates are well below those formerly used in agriculture (typically 10 kg or more a. i./ha/season).

3. Soils sampled to a depth of 5 cm contained 1–200 ppm ΣDDT dry weight beneath sprayed trees and <0.01–0.1 ppm ΣDDT dry weight at unsprayed sites in both sprayed and unsprayed areas.

4. Residues of DDT dissipated rapidly from mopane bark and soil by photolysis and volatilization. Deposits had half-lives of about 50 days on bark, 90 days in loamy sand and 125 days in silt loam. Fifteen to twenty-one percent of the initial deposit remained on the soil surface one year after treatment.

5. The high rates of dissipation under tropical conditions (half-life: 2–9 months) contrast with the slow rates observed in the temperate zone (half-life: 3–10 years).

Soil Fertility

6. High concentrations of DDT in soil can kill the soil fauna and damage soil fertility.

7. Decomposition of leaf litter, CO_2 production and nitrification were measured in soil to assess the impact of DDT on soil fertility.

8. Microbial activity, measured by litter degradation, CO_2 efflux and nitrification rates, was not affected by DDT residues. High residue concentrations reduced the degradation of litter by macro-invertebrates but the risk of a general reduction in soil fertility is low as high residue concentrations are restricted to the immediate vicinity of sprayed sites.

Invertebrates

9. Effects of DDT on invertebrate populations in, on, and above the soil were investigated both intensively at a single site, treated progressively over three years, and extensively at widely spaced sites in the sprayed and unsprayed areas.

10. Spatial and temporal variation in the abundance and diversity of invertebrates was high, possibly masking effects of DDT.

11. Surface-active invertebrates were sampled using a variety of methods, including pitfall traps. 72 400 animals were sorted, counted and identified from a trapping effort

---

* ΣDDT = DDT + DDD + DDE

of 2800 trap-days. Over 575 species were represented with many new to science. The dry season fauna was characteristic of an arid environment and dominated by Collembola (most abundant), spiders (most diverse) and insects, especially ants (greatest biomass).

12. At the intensive study site, biomass did not vary significantly between pitfall catches in sprayed and unsprayed sectors, but reduced abundance and species richness of certain taxa in the sprayed sectors may have been due to DDT.

13. Extensive sampling showed that there were significant faunal differences between sprayed and unsprayed areas, but whether this was due to spatial (habitat) variation or to DDT was unclear. At the intensive study site the fauna varied seasonally and annually, but minor effects of DDT could not be ruled out.

14. Relative abundance in many of the 95 most abundant taxa caught by pitfall traps varied inconsistently between sprayed and unsprayed areas. With the possible exception of certain mesostigmatid and prostigmatid mites and two ant species, *Ophthalmopane berthoudi* and *Platythyrea cribrinodis*, differences could not be attributed to DDT.

15. Soil-dwelling invertebrates were sampled using soil cores and litter bags. Over 14 000 animals were sorted and identified. Mites were the most abundant group in litter bag samples, whereas nematodes predominated in soil cores. About 100 species of mite and 22 species of nematode were identified.

16. Sprayed sector litter bags from the intensive study area contained fewer species of nematode and prostigmatid mite than bags from unsprayed sectors but species richness of other mite sub-orders was similar in sprayed and unsprayed sectors.

17. The relative abundance of nematodes and mites in soil cores did not vary with DDT treatment. However, the mite *Hypozetes* sp. (Cryptostigmata: Ceratozetidae) was more, and *Protogamasellus* sp. (Mesostigmata: Ascidae), less, abundant in sprayed sector litter bags than in unsprayed sector bags. These results are consistent with DDT-induced changes in temperate agricultural soils and are indicative of effects of ground-spraying on the invertebrate fauna in the study area.

18. Wooden baits were attacked more by termites in sprayed areas than in unsprayed areas.

19. Many invertebrates whose abundance appeared unaffected by DDT carried detectable residue burdens. Particularly high concentrations of ΣDDT (< 218 ppm dry weight) were found on some *Camponotus* spp. ants, which were a favoured food of the White-headed Black Chat.

Lizards

20. Surveys of species richness, relative abundance and population age structure were made in sprayed and unsprayed areas of mopane woodland and rock outcrop.

21. Nineteen species were recorded, 17 in woodland and 11 on outcrops. Species richness did not vary with spray treatment.

22. Numbers of the commonest skink, *Mabuya striata*, were lower in woodland treated more than twice (39% fewer sightings) and the proportion of adults sighted below 3 m

on tree trunks (i.e. the spray target) was also reduced. Residue concentrations in carcases increased with years of exposure and relatively high concentrations were found in lizards exposed to two or more treatments. It is concluded that the population of *Mabuya striata* is adversely affected by ground-spraying with DDT but that numbers should recover with the dissipation of residues.

23. No evidence of adverse effects or high residue burdens in other lizards was found.

Birds

24. DDT residues can kill, cause eggshell thinning and modify the behaviour of birds.

25. Over 300 bird species have been recorded from the study area; some are rare or of restricted range but none is endangered or endemic to Zimbabwe.

26. Effects on populations of the commoner woodland species and one bird of prey, the African Goshawk *Accipiter tachiro*, were investigated. The results of a separate study of residue concentrations and eggshell thinning in the Peregrine Falcon *Falco peregrinus* are also reported.

Common woodland birds

27. The relative abundance of common species was assessed by timed point counts at regularly spaced sites along tracks through sprayed and unsprayed areas at Siabuwa. More detailed surveys of three species, White-headed Black Chat *Thamnolaea arnoti*, Red-billed Wood-hoopoe *Phoeniculus purpureus* and White Helmet Shrike *Prionops plumatus*, were made using playback of tape-recorded calls and of one species, White-browed Sparrow-weaver *Plocepasser mahali*, by nest counts.

28. Species richness was similar in the sprayed and unsprayed areas but most species were less common in the sprayed area and 35–39% fewer birds were counted there.

29. Characteristics of the vegetation were recorded at the sample points and correlated with the relative abundance of 19 species. With the influence of 8 characterisics removed, 10 species were equally abundant in sprayed and unsprayed areas and 9 significantly less common in sprayed areas. The former group tended to feed more in tree canopy or less on arthropods than the latter, which fed on arthropods at or near the ground.

30. Closer study was made of DDT residue concentrations in seven species, Striped Kingfisher *Halcyon chelicuti*, Red-billed Wood-hoopoe and five songbird species, White-headed Black Chat, Black Tit *Parus niger*, Chinspot Batis *Batis molitor*, White Helmet Shrike and White-browed Sparrow-weaver, over spray treatment, time, feeding strategy and diet.

31. In sprayed areas, high residue concentrations were associated more with feeding site than diet. Highest concentrations were found in the wood-hoopoe and chat, which eat arthropods from tree trunks or from the ground. Lowest concentrations, about ten times less, were found in the flycatcher and helmet shrike, which feed on insects from tree and shrub canopy. High concentrations were also found in the sparrow-weaver which feeds on the ground but mainly eats seeds.

32. Residue concentrations were about ten times less a year after first being measured, 1–3 months after spraying. Residue concentrations in birds from unsprayed areas were about 100 times less than those found in birds from recently sprayed areas.

33. Surveys showed that chat and wood-hoopoe numbers fell by about 90% in an area subjected to three annual treatments with no comparable decline in an adjacent unsprayed area. None of the other five species declined significantly in the sprayed area. Chats bred successfully in sprayed areas and there was no evidence of adverse effects of DDT on their food supply. It was concluded that both chats and wood-hoopoes are at risk from accumulating lethal concentrations of DDT residues following DDT treatments. Surveys of chat numbers over a wide part of north west Zimbabwe indicate numbers have been depleted by ground-spraying with DDT.

34. Numbers of several other common woodland birds, including Fork-tailed Drongo *Dicrurus adsimilis*, Red-billed Hornbill *Tockus erythrorhynchus*, Arrow-marked Babbler *Turdoides jardineii*, Red-capped Robin *Cossypha natalensis*, woodpeckers (4 species) and doves (3 species) may have been reduced by ground-spraying.

35. The population dynamics of affected species in the area are not known, but given the rapid decline of residue concentrations in most species the prognosis for recovery is good. In the White-headed Black Chat and Red-billed Wood-hoopoe it may be slow and perhaps in the order of 10 years.

African Goshawk

36. Searches for nests were made along streams at Siabuwa during the breeding seasons of 1988/89, 1989/90 and 1990/91. Single eggs, from seven nests, were taken under licence to relate eggshell thickness to residue content and spray treatment. Eggshell fragments were taken from three more nests to relate shell thickness to spray treatment.

37. Eggshell thickness was inversely related to DDE content, which increased with number of past spray treatments of the nest site. Cracked eggs were found at two sprayed sites. Mean DDE concentrations in four out of six clutches from sprayed areas exceeded 130 ppm dry weight, the highest critical level associated with eggshell thinning and population decline in other raptor species.

38. The small number of occupied nest sites and higher rate of site desertion in the sprayed area in recent years, compared with the unsprayed area, strongly suggest ground-spraying with DDT caused a population decline.

39. The population dynamics of the African Goshawk in the area are unknown and the rate of recovery is uncertain. By analogy with the related European Sparrowhawk *Accipiter nisus*, recovery may take 10 to 20 years depending upon the rate of decline in DDE concentrations.

Peregrine Falcon

40. An independent survey of DDE residues in Peregrine Falcon eggs was made in Zimbabwe in 1990. Residues were found in all 11 samples but only one, from an area treated six years earlier, contained sufficient DDE to cause significant eggshell thinning.

41. The number of recorded Peregrine eyries has increased since 1971 due to the survey work of the Zimbabwe Falconers' Club and there is no evidence to suggest declining numbers in Zimbabwe.

Nocturnal Mammals

42. The relative abundance of insectivorous bats was systematically sampled in both wet and dry seasons at comparable sites in unsprayed and sprayed areas with mist nets and harp traps. A total of 1027 individuals were captured, catch effort being evenly distributed between treated and untreated sites.

43. The foraging activity of this fauna was also assessed with a bat detector along replicate transects on tracks through both untreated and sprayed woodland during 11 consecutive months. Bats found in day roosts were captured with hand nets.

44. Thirty species were recorded, 26 in the sprayed area and 22 in the unsprayed area; 18 were common to both areas.

45. No bats were found with DDT concentrations in the brain sufficient to cause mortality. However, lethal concentrations in individuals of five species, *Rhinolophus hildebrandtii*, *Nycteris thebaica*, *Scotophilus borbonicus*, *Hipposideros caffer* and *Eptesicus somalicus*, were projected with a decline of fat reserves to 1% due to acute stress, as might occur during a drought. Estimated mortality ranged from 0% in the unsprayed area to 38–100% at sites sprayed 5–6 times, depending upon species and excluding adult females, which lost residues to juveniles during lactation.

46. Bats accumulate residues via three major routes; the adult diet, direct contact at roosts, and from their mother's milk, and are at risk from mortality or sub-lethal effects which reduce survival or reproductive success.

47. ΣDDT concentrations on insects sampled by light trap in an area sprayed three months previously varied from 0.9 ppm wet weight (Lepidoptera) to 2.9 ppm wet weight (Orthoptera: Ensifera). Such residue concentrations in their diets readily account for the observed bioaccumulation of DDT in sampled bats. Seasonal and habitat factors relating to the presence of water appeared to have an overiding influence on the availability of potential prey, sampled by Malaise traps, which did not vary consistently between sprayed and unsprayed areas.

48. Highest mortalities were projected for *R. hildebrandtii* and *N. thebaica* both of which roost in hollow trees that were treated with DDT.

49. DDT concentrations in adult females were related to reproductive state, being higher in non-reproducing and pregnant females than in those coming to the end of lactation, after transfer of residues to juveniles in milk.

50. With one exception, there was no evidence of reproductive failure in sprayed areas. A high level of failure was found at one colony of *R. hildebrandtii* in a baobab tree which had been directly treated with DDT.

51. Neither the sex ratio nor age structure of populations varied significantly between sprayed and unsprayed areas. However, the proportion of mature male *E. somalicus* in the sprayed area was half that in the unsprayed area, consistent with an effect of DDT.

52. Body weights of *E. somalicus*, *S. borbonicus* and *Nycticeius schlieffenii* did not vary significantly between sprayed and unsprayed areas.

53. With one exception, capture-recapture figures and residue concentrations suggest there was no significant population exchange between sprayed and unsprayed areas.

The proportion of unaltered DDT in ΣDDT in lactating female *E. somalicus* caught in the unsprayed area in November 1989 indicated recent exposure.

54. The relative abundance of bat species varied more between habitats and with season than between spray treatments.

55. There was also no consistent difference between unsprayed and sprayed woodland in the level of foraging activity, with variability between replicate transects often as high as between treatments.

56. Surveys of other nocturnal animals were carried out along transects with a spotlight. The relative abundance of bushbabies (probably all *Galago moholi*), genet cats (probably all *Genetta tigrina*) and nightjars (Caprimulgidae) did not vary between sprayed and unsprayed areas.

Aquatic Environment

Fish

57. The Zambezi river system contains no endemic fish species. The fauna of the middle Zambezi has been changed significantly by the building of a dam at Kariba and subsequent colonization and introductions. Not all species have benefited, but 63 species are now known from the area compared with 48 before construction of the dam. Lake Kariba supports an important fishery, based on the introduced sardine *Limnothrissa miodon*, which yields about 20 000 m t a year and the lake also draws many angling tourists.

58. DDT is acutely toxic to fish and at sub-lethal concentrations may significantly affect cell membranes and enzyme systems, causing permanent loss of function. Potentially, DDT may affect recruitment, growth and survival.

59. No fish kills attributable to DDT have been reported from Lake Kariba or its affluent rivers. Potential chronic effects were investigated using standard techniques relating growth and mortality rates for several species to DDT residue burdens.

60. With the exception of the Zambezi, rivers flowing into Lake Kariba are seasonal and their faunas impoverished. Effects of spraying on the fish of one watershed river, the Songu, were studied using seine netting.

61. Three common and two rare species were caught. The Sharptooth Catfish *Clarias gariepinus* was more abundant in the upstream unsprayed section whereas the Red-eye Labeo *Labeo cylindricus* and Linespotted Barb *Barbus lineomaculatus* were more abundant in the downstream sprayed section. There was no evidence of an effect of DDT on species composition of seine net catches.

62. ΣDDT concentrations in all three species were significantly higher in fish from the sprayed area than in fish from the unsprayed area. The proportion of unchanged DDT in the total residue burden was highest in Sharptooth Catfish and lowest in Red-eye Labeo. This probably reflects the feeding habits of the different species, as the predatory catfish is likely to accumulate high concentrations of unaltered DDT from contaminated invertebrate prey.

63. Growth of Red-eye Labeo was similar in sprayed and unsprayed sections of the river based both on ageing by scales and length-frequency data analysed by ELEFAN.

Growth of the 1984/85 cohort was faster than growth of cohorts from 1985/86 and 1986/87 due to heavier rainfall in the former year. Overall mortality rates were similar in sprayed and unsprayed areas.

64. The most highly contaminated population of Linespotted Barbs showed the highest mortality rates, but also grew the fastest and the possiblity that DDT-induced mortality reduced competition and enabled better growth for survivors was not ruled out.

65. DDT residues in the most heavily burdened Sharp-tooth Catfish from the sprayed section of the Songu were high enough to provide a substantial source of contamination for fish-eating birds.

66. On Lake Kariba, sampling with a fleet of gill nets was undertaken at five sites, three sprayed and two unsprayed, in the Ume estuary. Seventeen species were caught. Variations in catch composition were attributed to seasonal changes and habitat differences and no effect of DDT was detected at the family level.

67. ΣDDT concentrations in most species did not reflect the history of spraying operations nearby and were generally lower than in fish from contaminated sections of the Songu river. The highest residue concentrations in Tigerfish *Hydrocynus forskahlii* were found in the larger river mouths where mature fish gathered to spawn. Residue concentrations in the Zambezi Parrotfish *Hippopotamyrus discorhynchus* were significantly higher at one sprayed site than at the other four sites. Brown Squeaker *Synodontis zambezensis* and Sharp-tooth Catfish from sites treated once were significantly more contaminated than samples from untreated sites but those from a site treated twice were intermediate. Butter Catfish *Schilbe mystus* at sprayed sites were significantly more contaminated than at unsprayed sites.

68. The proportions of DDT and its metabolites in ΣDDT varied between species but not between sites. Butter Catfish had the lowest proportion of unchanged DDT and Brown Squeaker the highest. There was no obvious correlation with diet.

69. Growth and mortality parameters were examined for Butter Catfish and Zambezi Parrotfish, as samples from sprayed sites were significantly more contaminated than at unsprayed sites. There was no evidence of an effect of DDT on the growth performance of either species.

70. The mortality rate of Butter Catfish at one sprayed site (S1) was lower than elsewhere and was associated with high DDT residue burdens. Mortality rates at the other four sites were similar although fish from the sprayed sites were heavily contaminated with ΣDDT.

71. Zambezi Parrotfish had lower mortality rates at one sprayed site (S3), where DDT residue burdens were highest. The possibility that DDT reduced mortality requires further investigation.

72. There was no evidence of an effect of DDT on fecundity of the Zambezi Parrotfish, with no differences in the number of eggs present in fish from treated and untreated sites. However, residue burdens in eggs from the sprayed sites were high enough to put fry survival at risk.

73. No evidence of major impacts of DDT on the fish fauna in the study areas was detected. There were no consistent effects on population parameters of the fish studied

and at geometric mean ΣDDT residue burdens up to 20 ppm lipid, no adverse effects on growth and mortality of any species appeared. At 60 ppm ΣDDT, there was a possibility of increased mortality in the Linespotted Barb, accompanied by increased growth rates amongst survivors. The implications of this for the population dynamics of the fish fauna of the area are unknown, but it is likely that effects will be minor if they occur at all.

74. There was no evidence to suggest ΣDDT concentrations in lipid were higher in fish than in invertebrates but residue concentrations in Fish Eagle eggs were substantially higher than in fish samples from the same area.

Birds

Fish Eagle

75. Low-level aerial surveys along the southern shore of Lake Kariba were made in 1987 to locate nests and monitor hatching success. The surveys were repeated in 1990 in the easternmost lake basin. Twenty clutches of eggs were collected under licence in 1989 and 1990 to measure eggshell thickness and residue concentrations in the contents.

76. Overall, over 72% of clutches hatched, but success was less than 50% at the eastern end of the lake, around Charara and Msango.

77. DDT and its metabolites were found in every egg. Generally, ΣDDT concentrations varied from 14–49 ppm dry weight per clutch, but 113–223 ppm dry weight were found in clutches from the eastern end of the lake and in one clutch from the mouth of the Sengwa river. The proportion of unaltered DDT in ΣDDT was significantly higher in areas recently ground-sprayed than in clutches from areas not treated for over six years, implicating tsetse control operations as the source of contamination.

78. ΣDDT and DDE concentrations were significantly correlated with the Ratcliffe index of shell thickness. Comparison with the indices of eggs collected before 1945 showed that the Ratcliffe index had declined by about 11% since the advent of DDT, but eggs from Charara and Msango were more than 20% thinner.

79. DDE concentrations and eggshell thinning at the eastern end of the lake were similar to those found in declining populations of the related North American Bald Eagle *Haliaeetus leucocephalus* and European White-tailed Sea Eagle *H. albicilla*, and there can be little doubt that poor hatching success around Charara and Msango was due to DDE residues. However, Fish Eagle nest densities in the area were higher than anywhere else on the lake.

80. Comparison with 1980 data showed residue concentrations had fallen in areas not ground-sprayed since then, but had increased in areas treated more recently.

81. It is concluded that DDT residues have reduced hatching success locally on Lake Kariba, but are not limiting population size. However, the Fish Eagle population may be declining due to increased settlement of the lake shore, resulting in the loss of safe nest sites and increased human predation of chicks. This problem is exacerbated by the destruction of nest sites in national parks and safari areas by unmanaged elephant populations.

Reed Cormorant

82. Residue concentrations of 22 organochlorine compounds in liver and visceral fat of 86 birds shot between January and October 1986 at the eastern end of Lake Kariba were analysed.

83. Residues of hexachlorobenzene, HCH, lindane, DDT and its metabolites were detected but only DDE was found in significant amounts.

84. ΣDDT concentrations varied seasonally, with highest concentrations recorded between May and September. These posed no significant risk to survival of full-grown birds but the visceral fat of adult females contained DDE concentrations associated elsewhere with eggshell thinning and breeding failure in a related species.

General Conclusions

85. Generally, wildlife populations were affected less by DDT than by seasonal changes in climate and spatial variation in habitat. However, adverse landscape scale effects were found in populations of four bird and one lizard species. Residues posed a significant risk to the survival of at least five species of bats under drought conditions. The comparative scarcity of several bird and terrestrial invertebrate species in sprayed areas was probably also due to DDT residues. There was no evidence of significant effects on soil fertility or on the relative abundance, growth and mortality of fish. Effects on the majority of less common species cannot be ruled out.

86. The effects of DDT treatments are reversible as residues dissipate from both the physical and biotic environments. Exposure patterns and the population dynamics of affected species are not sufficiently well known to assess recovery time, but based on temperate zone experience this may take 10–20 years.

87. The relatively low application rate of DDT, combined with a high dissipation rate of residues from bark and soil, may reduce the long-term risk to wildlife. However, the high rate of dissipation may increase the risk of acute poisoning.

RISK TO HUMAN HEALTH

1. The World Health Organisation (WHO) considers DDT safe to people when applied in the home for mosquito control, but there has been little conclusive research on the chronic effects of exposure to support this claim.

2. Residue concentrations in the breast milk of some Zimbabwean women are amongst the highest recorded. High residue levels pose no known risk to adults, but DDE concentrations in breast milk commonly exceed those causing hyporeflexia in infants[20]. The implications for child development are unknown.

3. Ground-spraying with DDT contaminates foodstuffs and is the major source of DDT residues found in fish from Lake Kariba[14]. However, ΣDDT concentrations in fish do not exceed either the Maximum Regulatory Limit (MRL) or Acceptable Daily Intake (ADI) set jointly by the WHO and Food and Agriculture Organisation (FAO).

4. The contribution of tsetse control operations to residue burdens in the general population is likely to be much less than that due to the domestic use of DDT for mosquito control.

Economic Assessment

1. Environmental damage costs to food resources, human health, recreation, ecosystem functioning and existence values were considered in economic terms.

2. No significant costs were identified in the study areas. However, costs depend upon the scale and place of treatment. Potentially, ground-spraying with DDT may affect the amenity value of tourist areas with major wildlife concentrations, such as national parks. The impact on the quantity and quality of tourist experiences in north-west Zimbabwe, though, is likely to be negligible.

Mitigation Measures

The adverse effects of ground-spraying with DDT can be reduced either by substituting an alternative insecticide, or by adopting an alternative control technique.

Alternative insecticide

In laboratory studies, three synthetic pyrethroid insecticides were identified as potential substitutes for DDT[21]. In 1989 and 1990, Tsetse Control Branch carried out ground-spraying trials with one, deltamethrin, to investigate its potential further. The 1989 trial covered 600 km² of the Omay Communal Area previously treated with DDT in 1988, while the 1990 trial covered 467 km² of land nearby, treated with DDT in 1988 and 1989. Assessment of the effects of the first trial on non-target animals was limited (Hustler *in litt.*), but a more extensive programme was undertaken between July and September 1990 by a team from NRI[22] (Chapter 12).

Deltamethrin is more toxic than DDT to crustacea and fish but the risk of exposure in these groups is low, provided spraying activities near water are closely supervised. Deltamethrin may also affect terrestrial arthropods more severely than DDT, and the ecological consequences of this require further study. However, unlike DDT, deltamethrin does not bio-accumulate and it degrades relatively quickly in vertebrates. While short-lived indirect effects on insectivorous vertebrates, through food depletion, are a possibility, deltamethrin should not harm vertebrates directly.

Tsetse populations were effectively controlled by deltamethrin but, due to the high cost of insecticide, the trials cost almost twice as much per unit area as concurrent spraying operations with DDT (J. Barrett, personal communication 1992). The costs, efficacy and safety of other pyrethroids, particularly alphacypermethrin and lambda-cyhalothrin[21], should now be examined in field trials.

Alternative control techniques

Two alternative control techniques, one involving aerial spraying and the other odour-baited, insecticide-impregnated black cloth 'targets', have been used in Zimbabwe in recent years. Both have fewer side-effects than ground-spraying[19].

Aerial spraying

Large areas of flat ground have been treated with fine mists of non-persistent insecticide, put down at night from low-flying aircraft (Plate 3). Treatment is indiscriminate but only low doses of insecticide, applied five times over a 2–3 month period, are needed. Remarkably, the proper application of endosulfan drift sprays kills no other animals except tsetse flies and some small fish, and the impact on the standing crop of fish is probably insignificant. Trials using deltamethrin as a substitute for endosulfan show it to be safer for fish, but more dangerous to tree canopy and aquatic arthropods. The main drawback to aerial spraying is its high cost relative to ground-based control techniques and its doubtful effectiveness in broken terrain (J. Barrett, personal communication 1992). In particular, it does not work against *Glossina pallidipes* at the dose rates used at present.

Targets

Since 1985 tsetse-attractive, cloth 'targets' impregnated with insecticide have become the mainstay of tsetse control in Zimbabwe (Plate 4). Over 70 000 targets are now deployed, mainly in barriers to prevent the re-invasion of areas cleared by ground- and aerial spraying. The ability to eradicate flies from large areas using targets alone remains in doubt. Theft, maintenance and the availability of materials can be problematic, but targets have minimal direct impact on non-target fauna and are one of the cheapest tsetse control options (J. Barrett, personal communication 1992). However, as with ground-spraying operations, service roads are necessary and construction of these may cause soil erosion and allow poachers access to game reserves. As a result, aerial spraying is the option preferred by the Department of Wild Life and National Parks for the removal of tsetse flies from Matusadona National Park[23].

CONCLUSIONS

The success of ground-spraying operations involving DDT is unrivalled. Over 50 000 km^2 of land has been cleared of tsetse, most with one or two treatments. This study shows that several bird and one lizard species have been severely affected by spraying. A wide range of other species may also be affected, but the effects of DDT are generally less than those of seasonal changes in the climate and habitat variation. Based on temperate zone experience, the prognosis for full recovery within 10–20 years of an end to spraying seems good.

The direct impact of ground-spraying with DDT in the Zambezi valley is trivial compared with the effects of the dam at Kariba or of the immigrant farmers, unmanaged elephant herds and frequent bush fires which have largely destroyed the mature miombo and mopane woodlands. This destruction is irreversible for the foreseeable future, even within the national parks, where resources are inadequate for effective management. To-day, habitat loss rather than pesticide pollution is the main threat to wildlife in the Zambezi valley.

In 1986 adequate national park management in Africa was estimated to cost at least US$ 200/km^2 a year[24], which at 1990 prices equals 533 Zw$/km^2, or almost exactly the extra cost of substituting deltamethrin for DDT in the ground-spraying trials of 1989 and 1990 (J. Barrett, personal communication 1992). The choice is therefore clear. If ground-spraying is retained as a control technique and DDT remains much cheaper than deltamethrin, DDT should be used and the savings invested in a compensatory project to protect wildlife habitat rather than to safeguard a handful of species through the introduction of deltamethrin.

In reality, the choice may not be so simple. The responsibilities for tsetse fly control and wildlife protection lie with separate government ministries, which traditionally compete rather than collaborate in bidding for financial resources. Tsetse Control Branch is committed to the adoption of safe control technologies and has announced it will phase out the use of DDT by 1995. It also wishes to retain ground-spraying as a control technique. However, in the absence of a cost-effective substitute for DDT, any decision to resume ground-spraying after 1995 will not be in the best interests of either the environment or the taxpayer unless government enables departmental collaboration to optimise environmental protection.

PROJECT PUBLICATIONS

Published or in press at December 1993

Disney, R.H.L. (1990) Revision of the Alamirinae (Diptera: Phoridae). *Systematic Entomology*, **15**: 305–320

Disney, R.H.L. (1991) Scuttle flies from Zimbabwe (Diptera, Phoridae) with the description of five new species. *Journal of African Zoology*, **105**: 27–48

Disney, R.H.L. (1992) The 'missing' males of the Thaumatoxeninae (Diptera: Phoridae). *Systematic Entomology*, **17**: 55–58

Douthwaite, R.J. (1992) Effects of DDT on the Fish Eagle *Haliaeetus vocifer* population of Lake Kariba in Zimbabwe. *Ibis*, **134**: 250–258

Douthwaite, R.J. (1992) Effects of DDT treatments applied for tsetse fly control on White-browed Sparrow-weaver (*Plocepasser mahali*) populations in NW Zimbabwe. *African Journal of Ecology*, **30**: 233–244.

Douthwaite, R.J. (1992) Effects of DDT treatments applied for tsetse fly control on White-headed Black Chat (*Thamnolaea arnoti*) populations in Zimbabwe. Part I: populations changes. *Ecotoxicology*, **1**: 17–30.

Douthwaite, R.J. (1993) Effects on birds of DDT applied for tsetse fly control in Zimbabwe. In: *Birds and the African Environment* (Wilson, R.T., ed.), *Proceedings of the Eighth Pan-African Ornithological Congress. Annales Musee Royal de l'Afrique Centrale (Zoologie)*, **268**: 608–610.

Douthwaite, R.J. (In press) Occurrence and consequences of DDT residues in woodland birds following tsetse fly spraying operations in NW Zimbabwe. *Journal of Applied Ecology*.

Douthwaite, R.J. and Tingle, C.C.D. (1992) Effects of DDT treatments applied for tsetse fly control on White-headed Black Chat (*Thamnolaea arnoti*) populations in Zimbabwe. Part II: Cause of decline. *Ecotoxicology*, **1**: 101–115.

Douthwaite, R.J., Hustler, C., Kreuger, P. and Renzoni, A. (1992) DDT residues and mercury levels in Reed Cormorants on Lake Kariba: a hazard assessment. *Ostrich*, **63**: 123–127.

Hartley, R.R. and Douthwaite, R.J. (1994) DDT residues and the effect of DDT spraying to control tsetse flies on an African Goshawk population in the Zambezi Valley, Zimbabwe. *African Journal of Ecology*, **32**: 000–000.

Lambert, M.R.K. (1992) Lizards, DDT and deltamethrin in NW Zimbabwe. In: *Proc. 6th Ordinary General Meeting, Societas European Herpetologica*, 19–23 August 1991, Hungary. Budapest: Hungarian Natural History Museum.

Lambert, M.R.K. (1993) Effects of DDT ground-spraying against tsetse flies on lizards in Zimbabwe. *Environmental Pollution*, **82**: 231–237.

Lambert, M.R.K. (1994) Ground-spray treatment with deltamethrin against tsetse flies in NW Zimbabwe has little short-term effect on lizards. *Bulletin of Environmental Contamination and Toxicology*, **53**(4): 000–000.

Tingle, C.C.D. (1993) Bait location by ground-foraging ants (Hymenoptera: Formicidae) in mopane woodland selectively sprayed to control tsetse fly (Diptera: Glossinidae) in Zimbabwe. *Bulletin of Entomological Research*, **83**: 259–265.

Tingle, C.C.D. (In press) The effects of DDT used to control tsetse fly on woodland invertebrates in Zimbabwe. In: *Proc. 11th International Colloquium on Soil Zoology*, 10–14 August 1992, Finland. *Acta Zoologica Fennica*.

Tingle, C.C.D. and Grant, I.F. (In press) The effect of DDT on litter decomposition and soil fauna in semi-arid woodland. In: *Proc. 11th International Colloquium on Soil Zoology*, 10–14 August 1992, Finland. *Acta Zoologica Fennica*.

Tingle, C.C.D., Lauer, S. and Armstrong, G. (1992) Dry season, epigeal invertebrate fauna of mopane woodland in northwestern Zimbabwe. *Journal of Arid Environments*, **23**: 397–414.

REFERENCES

1. Dunlap, T.R. (1981) *DDT: Scientists, Citizens and Public Policy*. Princeton: Princeton University Press.

2. Editorial (1971) DDT may be good for people. *Nature*, **233**.

3. Rhodesia (1972) *The Hazardous Substances Act*. Harare: Government Printer.

4. Zimbabwe (1985) *Statutory Instrument 220*. Harare: Government Printer.

5. Billing, K.J. and Phelps, R.J. (1972) Records of chlorinated hydrocarbon pesticide levels from animals in Rhodesia. *Proceedings and Transactions Rhodesia Scientific Association*, **55**: 6–9.

6. Whitwell, A.C., Phelps, R.J. and Thomson, W.R. (1974) Further records of chlorinated hydrocarbon pesticide residues in Rhodesia. *Arnoldia Rhodesia*, **6**(37): 1–8.

7. Greichus, Y.A., Greichus, A., Draayer, H.A. and Marshall, B. (1978) Insecticides, polychlorinated biphenyls and metals in African lake ecosystems. II. Lake McIlwaine, Rhodesia. *Bulletin of Environmental Contamination and Toxicology*, **19**: 444–453.

8. Wessels, C.L., Tannock, J., Blake, D. and Phelps, R.J. (1980) Chlorinated hydrocarbon insecticide residues in *Crocodilus niloticus* Laurentius eggs from Lake Kariba. *Transactions Zimbabwe Scientific Association*, **60**: 11–17.

9. Thomson, W.R. (1981) *A Report on Chemical Contamination of the Environment in Zimbabwe*. mimeo. p. 11. Harare: Department of National Parks and Wild Life Management.

10. Tannock, J., Howells, W.W. and Phelps, R.J. (1983) Chlorinated hydrocarbon pesticide residues in eggs of some birds in Zimbabwe. *Environmental Pollution Ser. B*, **5**: 147–155.

11. Thomson, W.R. (1984) DDT in Zimbabwe. In: *Proc. II Symposium on African Predatory Birds* (Mendelsohn, J.M. and Sapsford, C.W., eds) pp. 169–171. Durban: Natal Bird Club.

12. Thomson, W.R. (1981) Letter *Herald*, 30 April 1981, Harare.

13. Matthiessen, P. (1984) Environmental contamination with DDT in western Zimbabwe in relation to tsetse fly control operations. *Final Report of the DDT Monitoring Project*. pp. 90. Chatham: Overseas Development Natural Resources Institute.

14. Matthiessen, P. (1985) Contamination of wildlife with DDT insecticide residues in relation to tsetse fly control operations in Zimbabwe. *Environmental Pollution Ser. B*, **10**: 189–211.

15. Allsopp, R. and Hursey, B.S. (1986) *Integrated Chemical Control of Tsetse Flies (Glossina spp.) in Western Zimbabwe 1984–1985*. Harare: Irwin Press, for the Tsetse and Trypanosomiasis Control Branch.

16. Fox, P.J. and Matthiessen, P. (1982) Acute toxicity to fish of low-dose aerosol applications of endosulfan to control tsetse fly in the Okavango Delta, Botswana. *Environmental Pollution Ser. A*, **27**: 129–142.

17. Douthwaite, R.J. (1982) Changes in Pied Kingfisher (*Ceryle rudis*) feeding related to endosulfan pollution from tsetse fly control operations in the Okavango Delta, Botswana. *Journal of Applied Ecology*, **19**: 133–141.

18. Douthwaite, R.J. (1986) Effects of drift sprays of endosulfan, applied for tsetse fly control, on breeding Little Bee-eaters in Somalia. *Environmental Pollution Ser. A*, **41**: 11–22.

19. Douthwaite, R.J. (1992) Non-target effects of insecticides used in tsetse control operations. *FAO World Animal Review*, **70–71**: 8–14.

20. Rogan, W.J., Gladen, B.C., McKinney, J.D., Carreras, N., Hardy, P., Thullen, M., Tinglestad, J. and Tully, M. (1986) Neonatal effects of transplacental exposure to PCBs and DDE. *Journal of Pediatrics*, **109**: 335–341.

21. Holloway, M.T.P. (1989) Alternatives to DDT for use in ground spraying control operations against tsetse flies (Diptera: Glossinidae). *Transactions of the Zimbabwe Scientific Association*, **64**: 33–40.

22. Lambert, M.R.K., Grant, I.F., Smith, C.L., Tingle, C.C.D. and Douthwaite, R.J. (1991) Effects of deltamethrin ground-spraying on non-target wildlife. *Technical Report No. 1, Environmental Impact Assessment of Ground-Spraying Operations against Tsetse Flies In Zimbabwe*. Chatham: Natural Resources Institute.

23. Coulson, I.M. (1991) Tsetse fly eradication in Matusadona National Park: Integrated environmental planning to reduce conflicts with conservation and tourism. Unpublished report of the Department of National Parks & Wild Life Management, Harare.

24. Bell, R.H.V. and Clarke, J.E. (1986) Funding and financial control. In: *Conservation and Wildlife Management in Africa*. pp. 543–555. (Bell, R.H.V. and McShane-Caluzi, E., eds) Washington D.C.: Office of Training and Support, U.S. Peace Corps, (cited in Cumming, D.H.M. (1990) *Wildlife Conservation in African Parks: Progress, Problems and Prescriptions*. Multispecies Animal Production Systems Project Paper No. 15. Harare: World Wide Fund for Nature.

Part 2
Scientific Report

Effects on Wildlife

Terrestrial Studies

1

DEPOSITION AND DISSIPATION

I F Grant

Spray Drift

DDT droplets were collected on water-sensitive papers placed on the ground around sprayed trees (Figure 1.1). Under a prevailing wind speed of 1.5 m/s the heaviest droplets fell immediately around the tree and most of the medium and fine droplets (100–400 μm VMD*) landed within two metres of the target (Figure 1.2). Gusts of wind up to 3.5 m/s carried some droplets up to 4 m from the target. The finest droplets (<100 μm VMD), vapour and dust can be carried much further, accounting for the contamination of remote, unsprayed areas[1]. Residues transported from the target by rain and animal activity add to residues in the near vicinity over a longer period of time.

Residue Levels on Bark and in Soil

DDT residues on mopane bark showed wide variation in sprayed areas, occasionally reaching very high levels. This reflects the contagious distribution of DDT resulting from the spraying technique. There was no evidence of accumulation of residues on trees sprayed more than once. Residue levels on bark in unsprayed areas were generally close to detection limits.

Samples of bark, from approximately 1 m above ground, were removed by hammering a hollow steel soil corer into the tree trunk to the depth of the cambium. Average ΣDDT† residues on mopane bark from sprayed areas ranged from 0.5–93 μg/cm², compared with <0.01–0.03 μg/cm² on bark from the unsprayed area.

Contamination of soil around sprayed trees could be high but generally, residue levels from both sprayed and unsprayed sites in the study area were at the low end of the

---

* Volume Median Diameter (VMD) is the diameter at which half the spray volume is in larger drops and half in smaller drops
† ΣDDT = DDT + DDD + DDE

DDT in the Tropics

Figure 1.1 Spray droplets turned the yellow collecting papers blue and could be sized and counted (*Source*: C.C.D. Tingle)

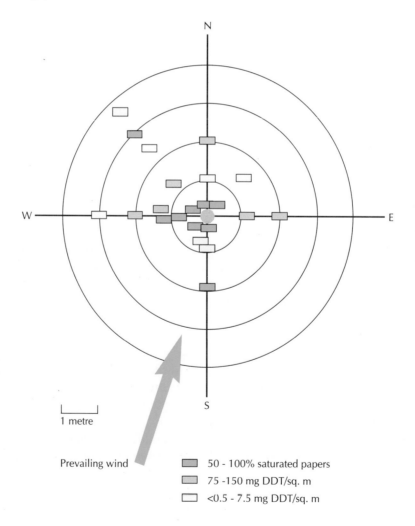

Figure 1.2 DDT deposits on the ground around sprayed trees

recorded range for natural land. Untreated soils from deserts, tundra and prairies typically contain 0.1 to 2 ppm ΣDDT dry weight, while agricultural soils may average 2 ppm, although this varies with the crop (<100 ppm in orchards) and cultural practice[2].

Soil samples to a depth of 5 cm were collected from a variety of sites and analysed for DDT residues. Samples from the unsprayed area contained between <0.01 and 0.1 ppm ΣDDT dry weight. Similar levels were found in samples from the sprayed area, unless the soil was taken below a sprayed tree. In the latter case, samples contained between 1 and 200 ppm ΣDDT dry weight.

DISSIPATION OF RESIDUES FROM BARK AND SOIL

Residues of DDT dissipated rapidly from mopane bark, having a half-life of about 50 days (Figure 1.3). Dissipation from soil surfaces was also rapid, with half-lives of about 90 days on loamy sand and 125 days on silt loam; however 15–21% of the initial deposit remained on the soil surface one year after treatment.

Dissipation over the first 100 days was not affected by rainfall, suggesting that dry season radiation levels promoted volatilization of DDT and its more volatile photolytic product DDE. Photolysis and volatilization can account for losses of 80–100% of DDT applied to tropical soil[3,4] and explain the relatively rapid dissipation of DDT under tropical conditions, in the order of 2–9 months as compared to 3–10 years in temperate zones[5,6,7].

The shorter half-life of ΣDDT on mopane bark than in soil is consistent with a greater opportunity for volatilization from bark, where the deposit is fully exposed to the air. The shorter half-life of ΣDDT in sandy as compared with silty soil is consistent with the relative adsorptive capacities of insecticides on sand and dry clay[2,8]. Compared with the work of Wessels *et al.* (unpublished) on the dissipation of DDT from cultivated soils in Zimbabwe[9], the half life of DDT at Siabuwa was roughly doubled. Wessels *et al.*, recorded half-lives of about 50 days in all but one of seven soil types, but rain had fallen at all sites within the first 100 days, and physical desorption and microbial degradation of ΣDDT usually increase with soil moisture. Moreover, cultivated soils have microbial populations adapted to pesticides, while woodland soils have more organic matter, perhaps the most significant factor influencing persistence[2].

REFERENCES

1. Cohen, J.M. and Pinkerton, C. (1966) Widespread translocation of pesticides by air transport and rain-out. *Advances in Chemistry Series*, **60**: 163–176.

2. Edwards, C.A. (1973) *Persistent Pesticides in the Environment*. Cleveland, Ohio: CRC Press.

3. Cliath, M.M. and Spencer, W.F. (1972) Dissipation of pesticides from soil by volitilisation of degradation products, 1. Lindane and DDT. *Environmental Science and Technology*, **6**: 910–914.

4. Yeadon, R. and Perfect, T.J. (1981) DDT residues in crop and soil resulting from application to cowpea *Vigna unguiculata* (L.) Walp. in the sub-humid tropics. *Environmental Pollution Ser. B*, **2**: 275–294.

5. Edwards, C.A. (1966) Insecticide residues in soils. *Residue Reviews*, **13**: 83–132.

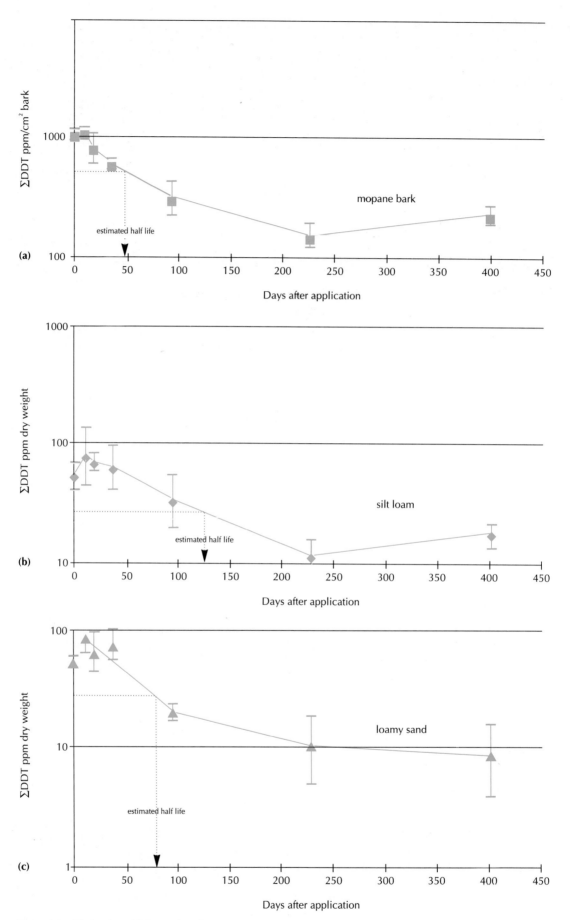

Figure 1.3 Dissipation of DDT residues from (a) mopane bark and (b,c) soil (geometric mean and standard error values shown)

6. FAO/IAEA (1986) *Report on the fate of persistent pesticides in the tropics, using isotope techniques.* Second FAO/IAEA Research Co-ordination Meeting, Quito, Ecuador 24–28 February 1986.

7. Hassan, A. (1991) Summary and Report of Second FAO/IAEA Research Co-ordination meeting on *'Behaviour of DDT in Tropical Environments'*, Jakarta, Indonesia 4–8 November 1991, Vienna: FAO/IAEA.

8. Wiese, I.H. and Basson, C.J. (1966) The degradation of some persistent chlorinated hydrycarbon insecticides applied to different soil types. *South African Journal of Agricultural Science*, **9**: 945–970.

9. Wessels, C.L., Tanock, J. and Phelps, R.J. (unpublished report) *Longevity of some chlorinated hydrocarbon insecticide residues on the surface of different soils.*

2

CRITICAL SOIL PROCESSES

I F Grant

DDT dissipates relatively quickly from surfaces and soils in tropical climates, but it remains a temporary hazard to surface-living and subterranean non-target organisms. By killing soil fauna and altering the community structure it can damage soil fertility in farmland[1–4]. Application rates in farmland were generally very high whereas contamination of soil in woodland sprayed with DDT for tsetse fly control is localized. Nonetheless, any effect on key degradative processes will increase the risk of reducing the productivity of soils already low in inherent fertility.

This chapter reports the results of three short-term studies designed to assess the impact of DDT on critical degradative systems that contribute to the maintenance of soil fertility. Rates of decomposition of leaf litter, soil respiration and nitrification were chosen as indicators of the breakdown and mineralization of organic matter.

LITTER DEGRADATION

The recycling of nutrients is essential for plant production, while soil organic matter influences soil fertility by promoting soil aggregation, increasing water holding and cation exchange capacities, and maintaining microbial biomass. The agents of recycling, micro-organisms and detritivores, are themselves links in food chains.

Dead mopane leaves were put in nylon mesh bags and buried or tethered to the ground in sprayed and unsprayed areas (Box 2.1, Figure 2.1). The bags varied in mesh size to allow different components of the decomposer community access to the leaves. Bags were reweighed after 6 or 17 months.

In the dry season, bags tethered on the surface lost little weight (mean 17% loss, all bags) irrespective of spray treatment or mesh size. However, the rate of litter degradation was significantly reduced at sprayed sites in buried bags accessible to macro-invertebrates. This effect was more marked in the loamy sands of *Julbernardia* woodland than in silt loam soil, despite more rapid dissipation of DDT residues from the former. There was no evidence DDT affected the breakdown of litter by small invertebrates and microbes, which processed as much as macro-invertebrates in unsprayed areas.

Figure 2.1 Mopane leaves were put in nylon mesh bags and buried in the ground to measure effects of spraying on decomposer organisms

Box 2.1 Litter decomposition

Nylon mesh bags filled with 3 g of dry, uncontaminated mopane leaves were buried to a depth of about 5 cm or tethered to the surface at various sites ($n = 12$) and left for 6 months (through one wet or one dry season) or for 17 months (two dry seasons and one wet). Several mesh sizes were used to allow access to different components of the decomposer community. In the dry season study, 64, 600 and 4000 μm mesh bags were used while 125 and 2000 μm mesh bags were used at other times.

Surface-tethered litter bags showed weight loss of less than 17% during the dry season irrespective of treatment or mesh size (Figure 2.2). This indicates that dry season microclimatic conditions (low humidity, high temperatures and radiation) at the woodland floor present detritivores and litter harvesters with an inhospitable environment. Nutrient wash-out from early wet season rains was probably the main agency of loss as found in other studies[8].

Dry season decomposition of litter buried in silty loam soil was most rapid in the largest mesh bags, where up to 60% was lost in 6 months (Figure 2.2). Decomposition rates in the 600 μm and 64 μm bags were significantly lower, between 30–38%, showing that subterranean macro-invertebrate decomposers were active in the dry season or during unseasonable early rains. There were no significant differences between decomposition rates in the smaller mesh bags or between rates at sprayed and unsprayed sites, suggesting that decomposition by microbes, mites and nematodes was unaffected by DDT. However, significantly more litter was removed from 4000 μm bags at the unsprayed site ($p = <0.5$) than at sites sprayed 1–3 times over a three year period suggesting the activity of subterranean macro-invertebrates was reduced by DDT during the dry season.

Litter degradation in buried, large mesh bags was faster in wet (85–90%) than dry season, but rates in the smaller mesh bags were comparable (35–38%) (Figures 2.2, 2.3). In silty loam soil losses during the wet season from 2000 μm and 125 μm mesh bags were similar at sprayed and unsprayed sites (Figure 2.3).

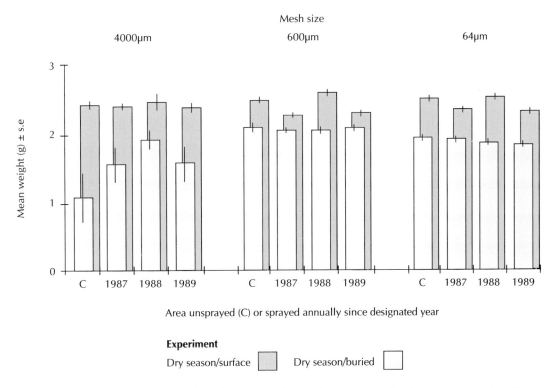

Figure 2.2 Leaf litter left in nylon mesh bags buried in silty loam soil and on the soil surface after six months of the dry season

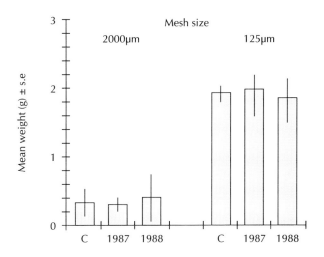

Figure 2.3 Leaf litter left in nylon mesh bags buried in silty loam soil after six months of wet season

Seventeen months after burial, 20–40% of the litter in 125 µm bags had decomposed, irrespective of soil type (Figure 2.4). Rates of decomposition at four unsprayed and five sprayed sites were similar, re-inforcing dry-season findings that DDT does not affect catabolic microbial activity. Over the same period, 89–99% of the litter in 2000 µm bags had been decomposed or removed in unsprayed and sprayed areas where soils were classified as silt loams. However, in loamy sands, supporting *Julbernardia* woodland, litter disappearance was significantly slower, and at three heavily contaminated sites it was reduced to 16–27%. Although the number of sites was limited, it appeared that not only were detritivorous macro-invertebrates less active or abundant in the sandy soil, but also that the effect of DDT on their activities were more severe in these soils than in silty loam.

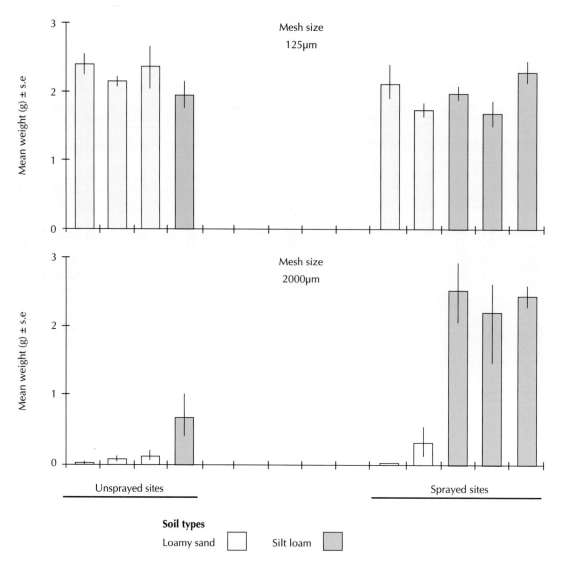

Figure 2.4 Leaf litter left in nylon mesh bags buried in silty loam and loamy sand soils after 17 months

Using termite (*Trinervitermes trinervoides*) bioassays, Wiese[13] demonstrated that soil texture markedly affected the adsorption and subsequent efficacy of DDT, which was only half as effective in a loam as in a sandy soil. The same bioassay showed a gradation in leaching of DDT. It is greatest in sand, less in loam and least in clay[14]. Hence, the retarded rates of litter processing by macro-invertebrates in the loamy sand might be explained by the effect of greater biological activity and downward movement of DDT in these soil types. Quantitative data on macro-invertebrate populations were not collected for sandy soils. Processing by microfauna and flora was unaffected by DDT, but overall rates of decay were slower in the loamy sand despite the more rapid dissipation of DDT residues. Fewer bound residues of DDT would be expected in a sandy soil[1], but as organic matter content in both soil types were comparable, low soil moisture was indicated as the constraint to microbial decomposition.

SOIL RESPIRATION

Carbon dioxide (CO_2) production 1 m from tree trunks was measured *in situ* at eight sites in the wet season and *ex situ* in samples of loamy sand and silty loam soils over 350–400 h using a portable infra-red gas analyser. *Ex situ* samples were mixed with dried grass (0.5% w/w), wetted and incubated to promote microbial and enzyme activity. Soil and air temperatures were recorded during measurement of CO_2 efflux to allow temperature-related fluctuations in respiration rates to be identified.

Ex situ respiration rates were generally higher in samples of silty loam soil than in loamy sand and no effects of contamination were apparent (Figure 2.5). *In situ* rates of CO_2 production varied quite widely and were generally higher in less contaminated soils (Table 2.1). However, *in situ* rates can be strongly influenced by root and animal activity. Local plant growth probably explains the highest respiration rate recorded, and low rates would be expected in two of the more contaminated samples on the basis of soil type alone. Although an effect of contamination cannot be ruled out, the natural variability of respiration within sites (CV* 20–50%) was not high enough to mask any serious suppression of heterotrophic activity, which has only been seen with very high dosage rates e.g. 50 kg/ha[5].

Table 2.1 *In situ* soil respiration at eight sites in sprayed and unsprayed woodland

ΣDDT residue ppm dry wt	Respiration rate mean (S.E.) (n=12) mg CO_2/m^2/h	Soil type
<0.01	211 (28)	Silty loam
<0.01	483 (34)	Silty loam
0.1	282 (19)	Silty loam
0.8	293 (28)	Silty loam
1.4	183 (17)	Loamy sand
1.7	189 (17)	Silty loam
2.6	198 (24)	Silty loam
39.4	203 (20)	Loamy sand

NITRIFICATION

Ammonium, produced from the mineralization of soil organic matter, is oxidized by nitrifying bacteria to nitrate. The availability of nitrate is critical for plant growth.

Nitrification rates in sandy clay loam soils were measured using the field method of Grant[6]. Samples were sieved to remove large objects and 100 g aliquots placed in plastic containers and amended with 100 µg ammonium/g dry weight of soil. Deionized water was then added to bring the soils to 70% of field capacity. Just before spraying, the containers were placed in pairs ($n = 3$) beside large trees in the spray area. One lid from each pair was removed and the soil exposed to spraying. After spraying, soils were incubated in the dark at ambient temperatures and nitrate accumulation measured at intervals for 50 days. DDT residues were measured in the samples at the end of the experiment.

There was no evidence nitrification rates were affected by spraying (Figure 2.6). About 60% conversion of ammonium to nitrate occurred within 27 days of exposure in both

---

* Coefficient of Variation (CV): the standard deviation as a percentage of the mean.

sprayed and unsprayed samples. Fifty days after spraying, sprayed soils contained mean residues of 8.1 µg ΣDDT/g dry weight of soil and unsprayed soils 0.1 µg ΣDDT/g dry weight of soil. Residue levels were probably at least 25% higher immediately after treatment.

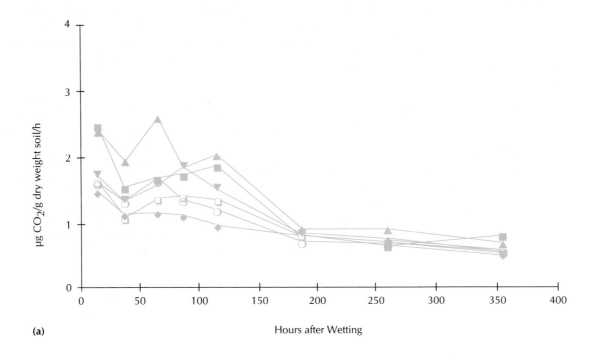

(a)

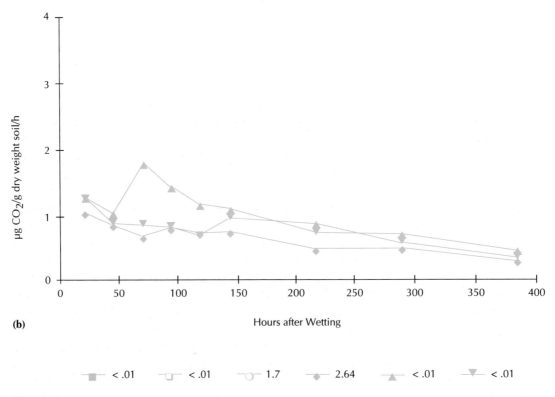

(b)

Figure 2.5 Respiration rates (µg CO_2/g dry wt. soil/h) of amended DDT-contaminated and uncontaminated (a) silty loam and (b) loamy sand soils

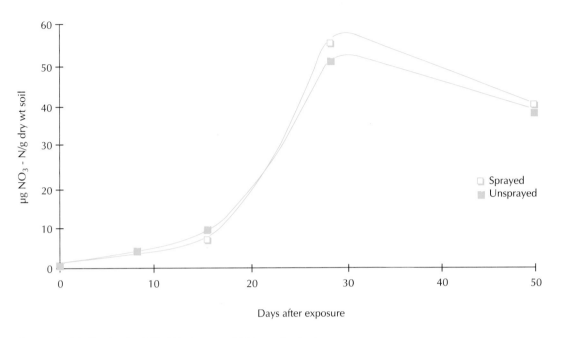

Figure 2.6 Nitrification (μg NO$_3$-N/g dry wt. soil/h) in sandy clay loam

Discussion

Plant litter is shed at the time ground-spraying operations with DDT are in progress. The small number of droplets missing targeted trees were rarely carried further than 5–10 m and the resulting distribution of DDT on litter and soil was, in general, contagious. After spraying, concentrations of ΣDDT in soil around trees were highly variable, averaging between 1–10 ppm; background levels were 0.01–0.03 ppm.

In the dry season, risks to decomposer organisms from DDT contamination were slight as microclimatic conditions (low humidity, high temperatures and radiation) on the woodland floor presented detritivores and litter harvesters with an inhospitable environment. The small losses from litter bags placed on the surface support this thesis. Losses may have been largely due to nutrient wash out by early rains.

Litter bags buried for six months in silty loam soil demonstrated that subterranean decomposers had been active in the dry season or during some unseasonal rain. Life underground was protected from direct contamination but indirect contact with DDT with the burial of contaminated litter and *via* percolation of rain water was assured. Decomposition by microbes (64 μm mesh) and microbes plus mites and nematodes (600 μm) was unaffected by DDT, but the macro-invertebrate fauna, which processed as much again, were significantly affected. As neither baiting for termite activity nor pitfall trapping revealed consistent differences in the relative abundance of surface-active invertebrates in sprayed and unsprayed areas (Chapter 3) it was concluded that subterranean macro-invertebrates were being affected. However, the impact of DDT on macro-invertebrates in silty loam soil was absent in the wet season, and as 90% of buried litter was processed within 6 months, it was concluded that decomposition overall was not impaired by spraying.

The effect of DDT on macro-invertebrates was exacerbated in the sandier soils. Quantitative data on macro-invertebrate populations were not collected, but it was clear that litter processing in *Julbernardia* woodland in the vicinity of sprayed trees had been seriously impaired by 18–51 ppm ΣDDT. Edwards[1] suggested that the greatest potential hazard of DDT residues was to the agents of litter processing in woodlands.

As the overriding factor governing the physical, biological and biochemical breakdown of organic matter is moisture, soil microbial and invertebrate activity and the related processes of soil respiration and nitrogen transformations are concentrated in the rainy season[7,8]. DDT is persistent and although its half-life on tree bark and soil is short by temperate standards (50 and 80 days), residues on mopane trunks are still considerable (viz. 20 µg ΣDDT/cm^2 bark) when the rains begin. Run-off from trees during the wet season increases soil DDT burdens nearby at a time when microbial activity, nutrient turnover and nitrogen requirements by young plants are at their highest. Quantification of the increase is difficult to estimate but wet season soil samples showed up to 500 µg ΣDDT/g dry weight soil (Figure 2.3).

In situ wet season respiration measurements made in the vicinity of sprayed and unsprayed trees failed to show differences attributable to DDT, despite soils containing up to 40 µg ΣDDT/g dry weight soil. The natural variability within sites (CV <50%) was not high enough to mask any serious suppression of heterotrophic activity, which has only been seen at very high dose rates e.g. 50 kg/ha[5]. Whether the CO_2 efflux truly reflected soil microbial activity is questionable with *in situ* measurements. However, *ex situ* monitoring in amended soils also provided no evidence of DDT disturbance. In all soils, the dynamics of decomposition of native organic matter and grass substrates were similar, suggesting that residues of DDT did not affect the overall microbial activity. Domsch *et al.*[9] argued that duration of pesticide stress on microbial activity is more important than magnitude, and speculated that depression of soil microbial activity was only detrimental when it exceeded 60 days.

Unlike many nitrogen transformations, nitrification is often suppressed by pesticides.[10–11] Residue levels of DDT in soils which were used in the nitrification study were typical of levels found around sprayed trees but fell short of concentrations shown to affect nitrification in temperate soils.[12] That no effect of DDT on nitrification was found suggests that assimilation of nitrate by plant communities was not impaired by ground-spraying for tsetse control.

Conclusions

Important soil microbial activities, including litter degradation, carbon and nitrogen mineralization, were unaffected by DDT. The activities of subterranean macro-invertebrates were temporarily affected by DDT in silty clay soils, but the overall function of the decomposer ecosystem appeared unimpaired. In some sandy loam soils, however, high levels of contamination caused significant medium-term reductions in litter degradation. Given that high residue levels are extremely localised the risk of a general reduction in soil fertility in *Julbernardia* woodland appears slight.

References

1. Edwards, C.A. (1973) *Persistent Pesticide in the Environment.* 2nd edn. Cleveland, Ohio: CRC Press.

2. Boyer, M.G. and Perry, E. (1973) Diversity in soil fungi as affected by DDT. *Mycopathologica et Mycologia applicata*, **49**: 255–262.

3. Cook, A.G., Critchley, B.R., Critchley, U., Perfect, T.J. and Yeadon, R. (1980) Effects of cultivation and DDT on earthworm activity in a forest soil in the sub-humid tropics *Journal of Applied Ecology*, **21**: 21–29.

4. Perfect, T.J., Cook, A.G., Critchley, B.R. and Russell-Smith, A. (1981) The effect of crop protection with DDT on the microarthropod population of a cultivated forest soil in the sub-humid tropics. *Pedobiologia*, **21**: 7–18.

5. Tate, K.R. (1974) Influence of four pesticide formulations on microbial processes in a New Zealand pasture soil. I. Respiratory activity. *New Zealand Journal of Agricultural Research*, **17**: 1–7.

6. Grant, I.F. (1988) Appropriate technology for monitoring environmental effects of insecticides in the tropics. pp. 147–156, In: *Field Methods for the Study of Environmental Effects of Pesticides.* (Greaves, M.P., Smith, B.D. and Greig-Smith, P.W., eds) British Crop Protection Council Monograph No. 40.

7. Swift, M.J., Heal, O.W. and Anderson, J.M. (1979) *Decomposition in Terrestrial Ecosystems.* Oxford, UK: Blackwell Scientific Publications.

8. Swift, M.J. Russell-Smith, A. and Perfect T.J. (1981) Decomposition and mineral-nutrient dynamics of plant litter in a regenerating bush-fallow in sub-humid tropical Nigeria. *Journal of Ecology*, **69**: 981–995.

9. Domsch, K.H., Jagnow, G. and Anderson, T. (1983) An ecological concept for the assessment of side-effects of agrochemicals on soil organisms. *Residue Reviews*, **86**: 65–105.

10. Atlas, R.M., Pramer, D. and Bartha, R. (1978) Asessment of pesticide effects on non-target soil organisms. *Soil Biology and Biochemistry*, **10**: 231–239.

11. Anderson, J.R. (1978) Pesticide effects on non-target soil microorganisms. pp. 313–533. In: *Pesticide Microbiology.* (Hill, I.R. and Wright, S.J.L., eds) London: Academic Press.

12. Ross, D.J. (1974) Influence of four pesticide formulations on microbial processes in a New Zealand pasture soil. II. Nitrogen mineralisation. *New Zealand Journal of Agricultural Research*, **17**: 9–17.

13. Wiese, I.H. (1964) Some biological studies on the inactivation of insecticides by various soil types. *South African Journal of Agricultural Research*, **7**: 823–836.

14. Wiese, I.H and Basson, C.J. (1966) The degradation of some persistent chlorinated hydrocarbon insecticides applied to different soil types. *South African Journal of Agricultural Research*, **9**: 945–970.

3

TERRESTRIAL INVERTEBRATES

C C D Tingle

INTRODUCTION

Invertebrates are extremely numerous and highly successful animals, which make many important contributions to the functioning of the living world. Some are involved in decomposition processes and nutrient recycling and some with pollination, whilst others have a major impact on plant biomass through herbivory or play a part in regulating animal populations through predation or parasitism. Invertebrates, in turn, provide the food for many vertebrates.

DDT was developed specifically to kill insect pests and so may be expected to have its most severe non-target impact on other insects and invertebrates. However, despite its broad spectrum of activity, relatively few invertebrate taxa have been shown to suffer adverse population effects from agricultural use of DDT, though some groups, particularly predatory, mesostigmatic mite populations, have been shown to decline after treatment.[1] Other soil mites may also be affected, particularly in the short term and in response to high doses of the insecticide.[2,3] This has knock-on effects on the normal prey of these mites, including some detritivorous mites [Cryptostigmata][4], nematode worms[5], and springtails [Collembola: Isotomidae].[6–8] Some species within all these groups are tolerant of DDT and increase in number following treatment.[9]

The majority of work on environmental impact of DDT has been done in farmland, but in a wide-ranging study of the effects of aerial spraying of DDT on forest invertebrates in the USA, a number of insect groups were found to be affected by the insecticide in the short term. Included in this were many bugs [Hemiptera], which were greatly reduced or completely eliminated, and several species of flies [Diptera] (especially Calypteratae). Wasps [Hymenoptera], from the superfamilies Prototrupoidea and Chalcidoidea, also declined in numbers after spraying, and several species of ant also seemed to be affected e.g. *Tapinoma sessile* and *Aphaenogaster treatae*.[10]

The ecological implications of these changes are unknown. In agro-ecosystems, the disruption caused by DDT to the natural balances within the system are recognized and have sometimes led to increased pest problems.[11] Undoubtedly, similar changes will occur in more complex, natural habitats, but are less easy to detect than in the simplified ecosystems occurring in agriculture and horticulture. Until the ecology of natural habitats is better understood, the long-term impact of pesticides used in these areas will remain unclear.

Invertebrates played an important role in the ecological impact of DDT in temperate areas, as a route for entry of the insecticide into the food chain. For example, earthworms surviving DDT applications tend to contain higher DDT residues than the surrounding environment, with the concentration factor averaging about 9-fold.[12–14] Contaminated worms were implicated in deaths of robins (*Turdus migratorius*) feeding around elm trees sprayed with DDT against vectors of Dutch Elm disease in Urbana, USA.[15] There is also evidence that invertebrates continue to pick up residues for considerable periods after spraying has ceased and even in areas that have been previously unsprayed, so providing a potential reserve for accumulation through food chains.[16]

The few attempts to monitor environmental impact of ground-spraying operations against tsetse fly[17] have paid little attention to invertebrates. There are reports of effects on non-target insects, but no further details are given.[18] The only specific example noted is the elimination of hersiliid spiders from the bark of sprayed trees in Chad.[19]

STUDY RATIONALE

Before the ecological effects of DDT use in the Siabuwa area could be assessed, baseline data on the composition of the invertebrate fauna were required. However, the sheer diversity of invertebrates precluded studying all of them and only the groups likely to be at greatest risk were investigated.

Study was restricted to the three groups at direct risk from DDT sprayed onto tree trunks, under fallen logs, under overhanging rocks and on the soil under thicket vegetation. These were epigeal invertebrates (those living on or above the soil surface); arboreal arthropods (specifically those living on and visiting tree trunks) and soil dwellers (also at risk from indirect contamination from wash-off).

There were a number of woodland habitat types in the study area, but invertebrate sampling was confined to *Colophospermum mopane*-dominated woodland. This is an important habitat for tsetse in the Zambezi valley and for ground-spraying operations.[20]

The use of widely spead sample sites allowed for any variation within treatment areas, but presented a number of problems in interpretation of the collected data, including lack of pre-spray information on sprayed sites and a variety of uncontrolled environmental factors within and between sprayed and unsprayed sites.

A spray monitoring area (SMA) was thus set up for more detailed study of the impacts of spray operations, to allow for pre- and post-spray sampling in a previously unsprayed habitat, something which could not be done over the rest of the study area (Box 3.1). This aided interpretation of the data from more widespread sites, where effects of the pesticide on invertebrate populations may otherwise have been overlaid by changes due to other environmental variables.

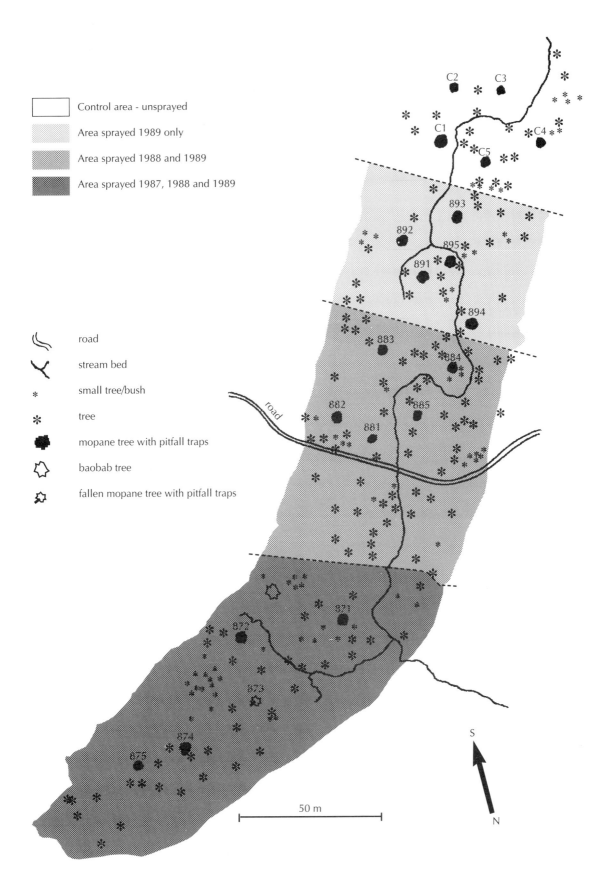

Figure 3.1 DDT impact assessment spray monitoring area showing invertebrate sampling points

Sample points were selected, where possible, to coincide with those used by other members of the study team, particularly the ornithologist and soil microbiologist. This allowed any changes in bird populations or in the functioning of soil processes to be related to invertebrates found at the same sites. Sampling was carried out during the dry seasons (June and July) of 1988, 1989 and 1990.

Box 3.1 Invertebrate monitoring

Invertebrates were monitored at 20 sites spread widely over the study area, in both sprayed and unsprayed habitat, using pitfall traps, termite baits, soil cores and ant nest surveys. Only a small proportion of a 'sprayed area' is actually treated with DDT during ground-spraying operations and targetting of the insecticide is very specific. Sample sites within the sprayed area (S1–S10) were restricted to the zone treated in consecutive years since 1984, and were selected by examining trees for visible DDT deposits remaining from the previous year's operation. This ensured that only fauna directly exposed to the insecticide were monitored, whereas random selection of sites in the sprayed area could have resulted in sites being some distance from the nearest sprayed tree. For many sedentary invertebrates, such sites would be, to all intents and purposes, unsprayed. Study sites in the sprayed area thus represent the worst case, reflecting the conditions directly within spray swaths, (approx. 20% of the total area). Selected sites in the sprayed area were then matched with 10 similar sites in the unsprayed area (U1-U10). A study site consisted of one tree within an area of mopane woodland.

The 'spray monitoring area' (SMA) was set up in an area of mopane woodland stretching across the 1987 spray boundary into the unsprayed zone, at a point where the projected 1988 spray boundary coincided with the 1987 boundary (Figure 3.1). The area was divided into four sections:

'87 area': sprayed in 1987 (with DDT deposits still visible on a number of trees in June 1988) and again in 1988 and 1989;

'88 area': previously unsprayed, sprayed for the first time in 1988 and again in 1989;

'89 area': previously unsprayed, sprayed for the first time in 1989;

'C area': unsprayed before and throughout the study.

Each area measured 50 m by approximately 40–60 m (the swath width of the spray team) and contained at least 25 trees of a size suitable as tsetse resting sites. Five of these were used as invertebrate sampling points for pitfall trapping[21], soil coring, termite baiting[22] and litter bag studies[23]; another five were used for trunk trapping. Ant baits were set up in the central part of each sector.

INVERTEBRATES OF MOPANE WOODLAND

Epigeal arthropods

A total of 72 400 epigeal invertebrates from 400 pitfall traps (trapping effort of 2800 trap days) were sorted, counted and identified. Over 575 species were represented (Box 3.2), with many new to science. Springtails (Collembola) were most abundant within the catch (Figure 3.2a), but the spider fauna showed the highest diversity. Insects were caught in large numbers, made up the highest proportion of the biomass (Figure 3.2b) and were dominated numerically by the ants. The faunal composition shows characteristics typical of arid environments.[21] The fauna represent a wide range of trophic levels including primary consumers, detritivores and scavengers, predators and parasites, all important in the functioning of the woodland ecosystem. In turn, they provide the major food source for a variety of reptiles, birds and mammals.

Terrestrial Invertebrates

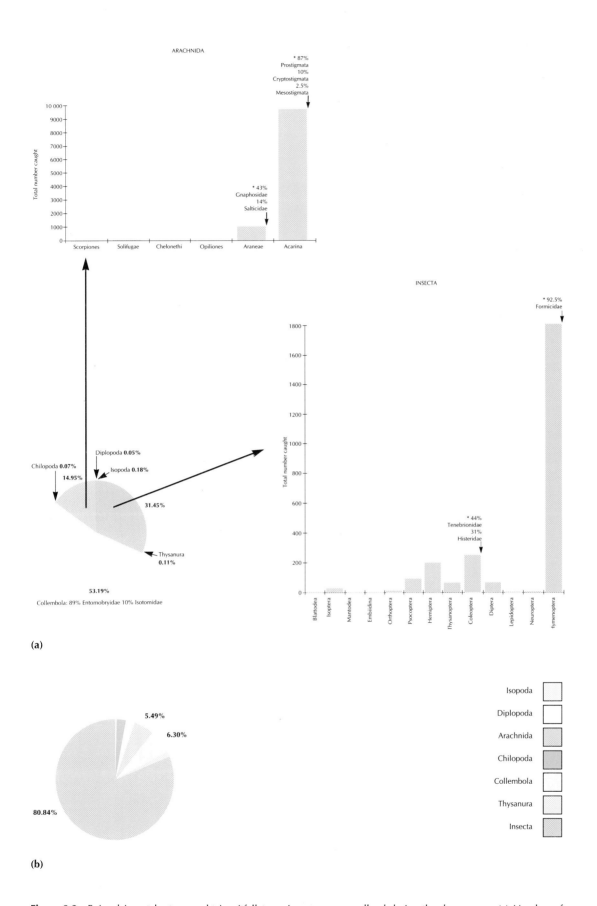

Figure 3.2 Epigeal invertebrates caught in pitfall traps in mopane woodland during the dry season (a) Number of individuals from 2800 trap days   (d) Biomass of individuals from 80 trap days

Box 3.2 Epigeal invertebrates

Epigeal invertebrates were sampled using pitfall traps (Figure 3.3), modified from the design described by Grant[23]. The invertebrates caught span 7 classes, 26 orders, 183 families and over 575 species. The general composition of this fauna is given in Figure 3.2. The invertebrate fauna trapped is of considerable taxonomic interest, with 27 species new to science in the collection.[25–27] There were probably many other undescribed species present as well, but specialist taxonomic assistance was not found to identify all groups.

(a)

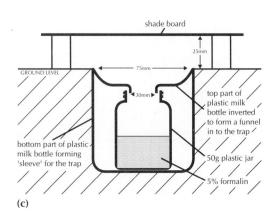

(b)

(c)

Figure 3.3 (a) Pitfall traps around mopane tree (b) Entrance to pitfall trap (c) Design of pitfall trap

The Isopoda, Chilopoda and Diplopoda are poorly represented (Figure 3.2). The Isopoda are omnivorous detritivores, the Diplopoda feed on a variety of dead plant material and all the Chilopoda are predators, although some also take decaying plant material.

Amongst the Arachnida, the Acarina were caught in the largest numbers (Figure 3.2). The Prostigmata predominate and include fungivores, bacterivores and predators (Figure 3.4). The detritivorous or fungivorous Cryptostigmata are next most numerous, with relatively few of the predatory Mesostigmata, the bacterivorous or fungivorous Astigmata and the ectoparasitic Metastigmata. Identification of mites was only attempted to family level. In general, it appears that the fauna was relatively limited consisting of only a handful of species per family.[28]

The 1078 spiders caught come from 25 families and show the highest diversity of all the epigeal

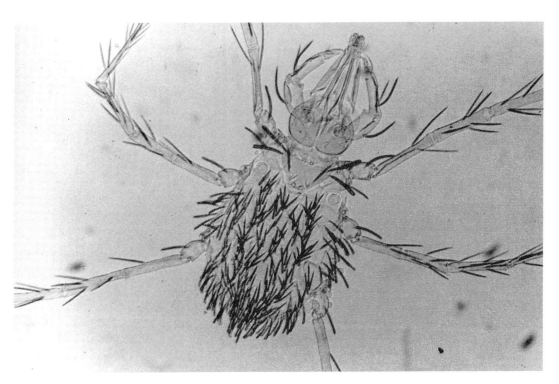

Figure 3.4 The predatory prostigmatid mites e.g. *Leptus* sp. (Erythracidae) were most numerous amongst the Arachnida, but the species richness of the group in sprayed areas is consistent with an impact of DDT (Source: C. Tingle; NRI)

invertebrates (Figure 3.5).[24] The Gnaphosidae and Salticidae are the most numerous and species-rich families. Classification of spiders to genus and species is tentative, as the majority caught in pitfall traps were immature and therefore difficult (or impossible) to identify past family level.[29] All are predators.

Within the Insecta, Hymenoptera and Coleoptera predominate. The ants are most numerous, and scavengers, seed harvesters and predators are all represented. The Myrmicinae are most abundant and species-rich, dominated numerically by various *Pheidole* species, which were common across the whole study area.

The Coleoptera were numerically dominated by two families: the Tenebrionidae and the Histeridae. The tenebrionids were largely from the sub-family Tentyriinae and are probably detritivorous as larvae and adults, whilst the histerids may be predators of small invertebrates.

Few Isoptera were caught in the pitfall traps, despite their importance in this habitat.[30] The harvester termite *Hodotermes mossambicus* (Hagen) [Hodotermitidae] was caught in the largest numbers, with small numbers of mound-building *Macrotermes* spp. and *Odontotermes* spp. [Termitidae].

The composition of the pitfall trap catches indicate a number of faunal characteristics typical of arid environments,[31] with Collembola outnumbering Acarina; larger numbers of Prostigmata than of Cryptostigmata; relatively low species richness of the ant fauna; presence of *Hodotermes mossambicus* and numerical dominance of Tenebrionidae within the Coleoptera.

Figure 3.5 The diversity of spiders (e.g. Jumping spiders (Salticidae) represented by 18 species) was the highest amongst the invertebrates and appeared unaffected by DDT (Source: Alan Weaving; Ardea)

Arboreal arthropods

Invertebrates walking up mopane tree trunks were sampled using trunk traps. The catch contained fewer species than the pitfall traps and comprised a mixture of specialist trunk dwellers, predominantly epigeal species and a number of species of flies (Diptera) (Figure 3.7). A total of 6811 invertebrates were trapped over 420 trap days, suggesting that there were fewer invertebrates on the tree trunks than walking on the soil surface (assuming equal efficiency of the traps). As with the epigeal fauna, the springtails and ants were the most numerous (Box 3.3).

Box 3.3 Arboreal arthropods

The traps used for sampling trunk fauna were based on a design developed in New Zealand.[31] The traps were made of locally available materials — a plastic sandwich box trap and a thick grade polythene collar around the tree trunk (Figure 3.6). The trap was generally set at about 1 m above the ground. The gap between the tree trunk and the plastic collar was sealed with mud to block crevices in the bark.

The faunal composition was limited by comparison with epigeal fauna caught in pitfall traps. Only three classes were represented, with 13 orders, 52 families and 75 species. As with the pitfall trap catches, Collembola were numerically dominant, comprising 90% of the catch (Figure 3.7).

The Arachnida were dominated numerically by mites, which showed a similar composition to the ground-dwelling fauna. The detritivorous Cryptostigmata were grossly outnumbered by Prostigmata, and predatory Mesostigmata occurred only in small numbers. The spider fauna of the tree trunks also shared many families with the ground-dwellers, although the species composition within families was somewhat different. Numerically, spiders made up an even smaller percentage of trunk trap catches than they did of pitfall trap catches.

Within the Insecta, the proportion of Diptera and Thysanoptera in the catch was notably greater in the trunk traps than the pitfalls, whilst the Isoptera were completely absent from trunk traps and Coleoptera reduced to very low numbers (Figure 3.7). Again the ants were numerically dominant, but the fauna was species poor by comparison with the pitfall trap catches.

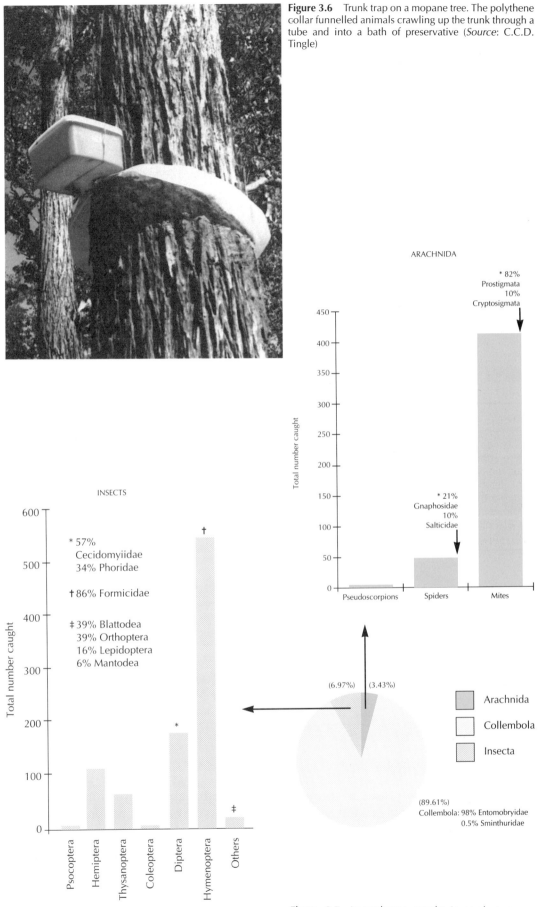

Figure 3.6 Trunk trap on a mopane tree. The polythene collar funnelled animals crawling up the trunk through a tube and into a bath of preservative (*Source*: C.C.D. Tingle)

Figure 3.7 Invertebrates caught in trunk traps on mopane trees during the dry season

Soil invertebrates

The sample of invertebrates extracted from soil cores indicates a very low density of soil invertebrates by comparison with temperate forest[32] (Table 3.1). Nematodes occurred in the largest numbers and there were relatively few micro- and macro-arthropods. No earthworms were recorded at all from soil samples taken in the dry season, nor using a standard formalin drench technique.[33] The absence of earthworms has been noted in previous studies in the drier tropics, where their role in nutrient cycling is frequently taken by termites.[34]

Box 3.4 Soil invertebrates

Invertebrates were extracted from soil cores taken to a depth of 5 cm, around the bases of sample trees, using a simple flotation technique. In general, the density of micro-arthropods was lower (Table 3.1) than in either temperate or African woodland[35], but was similar to those in a banana plantation in Uganda[36]. In that case, no figures were given for nematodes, but Collembolan density was comparable ($1.8 \pm 0.37 \times 10^3$/m), whilst mite populations appeared somewhat higher than in this study at $6.87 \pm 0.88 \times 10^3$/m.

Fauna were extracted from litter bags containing dry mopane leaves buried in the soil around sample trees in the SMA (Chapter 2). More micro-arthropods were recovered from litter bags than from the soil cores and nematodes were also well represented. Over 10 000 individuals were identified, representing 6 classes, 23 orders, 97 families and about 150 species. The mites were the most abundant group and were the best represented, with some 100 species. Nematodes were less numerous, but 22 species were identified. Relatively few springtails were found, mostly from the family Isotomidae and insect numbers varied greatly, but were generally low. Pot worms (Enchytraeidae) were also found in litter bags, though in small numbers.

Table 3.1 Estimated density of soil invertebrates in soil from the study area.

Group	Density (No./m²)
Nematoda	117.1×10^3
Micro-arthropods	5.2×10^3
Acarina	3.7×10^3
Collembola	1.4×10^3
Macro-arthropods	7.9×10^3
Diptera (larvae)	1.2×10^3
Coleoptera (adult)	0.4×10^3
Coleoptera (larvae)	0.4×10^3

THE EFFECTS OF DDT

Epigeal arthropods

The biomass, diversity, faunal composition and relative abundance of individual taxa within pitfall trap catches were compared between sprayed and unsprayed sites.

Biomass

Biomass was very variable within sectors of the SMA (Box 3.5) and there were no significant differences detected between sprayed and unsprayed areas (Table 3.2). There is thus no evidence that DDT affected the biomass of surface-active invertebrates in the period immediately following spraying, or in the long term.

Box 3.5 Biomass

During sorting of post-spray pitfall trap catches from the spray monitoring area for 1988, invertebrates were measured using an eye-piece graticule on a binocular microscope. The biomass of invertebrates was then estimated using a weight versus length relationship.[37]

For the majority of invertebrates weight was calculated using the formula:

$$W = 0.0305 \, L^{2.62}$$

However, for Diplopoda and Chilopoda a linear equation was used.[38]

$$W = -0.792 + 0.571 \, L$$

where W = weight and L = length.

The results in Table 3.2 show that biomass estimated in this way was similar in all four sectors of the SMA, whether sprayed or unsprayed. Variance of biomass estimates for catches in the two sprayed sectors was higher, but there was no evidence to suggest that DDT had any adverse impact on the biomass of surface-active invertebrates in the area.

Table 3.2 Invertebrate biomass collected in June 1988, 2 weeks post-spray

			Biomass (mg)	
Area		n	$\bar{X}$	s.e.
Sprayed	87	5	476.7	175.8
	88	5	536.3	123.8
Unsprayed	89	5	311.6	84.1
	C	5	329.8	59.4
Anova: F-ratio = 0.86		P = 0.5 n.s.		

n = number of sites (each with 4 pitfall traps)
$\bar{X}$ = mean
s.e. = standard error

Abundance and Diversity

Fewer invertebrates were caught in the '87' area than in any of the other sectors of the SMA throughout the period of the study (Figure 3.8). This was the most heavily sprayed sector of the SMA, but as there is no pre-spray data for this sector, the differences cannot be attributed conclusively to DDT. They may relate directly to inherent environmental differences in the area. However, there are indications that DDT may have been affecting invertebrates in the SMA, with fewer invertebrates caught in the '88' sector a year after it was sprayed for the first time, whilst numbers rose in both the unsprayed sectors.

Species richness of the pitfall trap catches shows a similar pattern (Figure 3.9). Whilst there was an overall decline in 'species' richness as the study progressed, this was most marked in the area sprayed for the first time during the study. The results do not constitute proof of a detrimental impact of DDT, although this seems likely. Certainly, other environmental variables are more important than spraying in determining the abundance and diversity of surface-active invertebrates, as neither the total number of invertebrates nor the number of taxa caught in pitfall traps showed differences between unsprayed (U) and sprayed (S) sites within the whole study area.

A variety of diversity indices were also calculated for the pitfall trap catch data and these show similar results with no evidence of differences in diversity between sprayed and unsprayed areas in the extensive survey, but indications of minor impacts of DDT from the intensive study in the SMA.[39]

DDT in the Tropics

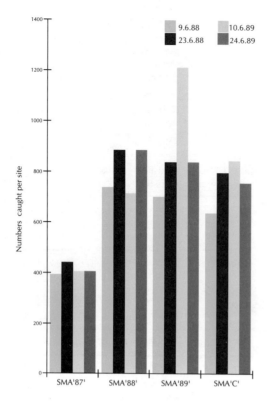

Figure 3.8 Mean number of individuals caught at sites within the spray monitoring area before and after spraying in 1988 and 1989

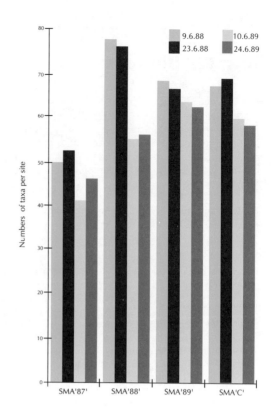

Figure 3.9 Mean number of taxa caught at sites within the spray monitoring area before and after spraying in 1988 and 1989

Similarity in Faunal Composition Different sites can be grouped using cluster analysis to show the similarity of their fauna (Box 3.6). Figure 3.10 shows clear separation of the sprayed (S) from unsprayed (U) sites in the extensive survey area. However, the lack of pre-spray information on these sites and their spatial separation leave the possibility that factors other than DDT may be responsible for the differences in faunal composition between the areas. Within the SMA there was no simple division into sprayed and unsprayed sites. Faunal similarity was governed principally by natural seasonal and annual variation, with some indications of an underlying long-term effect of DDT. There are indications that sites sprayed for the first time during the first year of the study (1988) became less similar to the unsprayed sites, a year after spraying. Thus, natural variation has a greater impact on faunal composition of sites than DDT, but there is evidence that DDT may have some effect.

Box 3.6 Faunal composition

Similarity of sites based on the invertebrate faunal composition of pitfall trap catches was quantified using Sorensen's Quotient of Similarity:

$$Q/S = (2j/a+b) \times 100$$

where j = number of species common to both sites;
a = number of species at site A;
b = number of species at site B.

The similarity quotients were then subjected to hierarchical, average link, cluster analysis. This procedure involves construction of a coincidence table comparing each site with every other one and selecting that pair of sites with the highest quotient of similarity. These are then grouped and quotients of similarity calculated between this group and the remaining sites. The procedure continues until the sites are classified into two groupings within which the extent of similarity between the individual sites is shown by the cluster arrangement. The results are then drawn up as a dendrogram.[32]

Figure 3.10a shows that, with the exception of two sites, faunal composition in the sprayed area is different from that in the unsprayed area. However, this pattern is not repeated in the SMA, where factors other than DDT are more important in determining faunal composition. Figures 3.10b and c show very different patterns from each other, indicating that short-term seasonal variation in faunal composition is outweighed by natural annual fluctuations. Certainly, no consistent effect of DDT is noticeable. Figure 3.10d shows that sites were primarily separated on the basis of sampling time, with all sites sampled in 1988 being more similar to each other in terms of their faunal composition (regardless of treatment) than the same sites sampled in 1989. In both years, the '87' area sites stand out as being different and form their own cluster. This may be due to an effect of DDT, as this is the most heavily sprayed part of the SMA, but may represent inherent differences in the fauna of this site. However, there were indications of a secondary effect of DDT in the longer term, underlying the primary grouping of sites on the basis of sample date (Figure 3.10d). Thus the '87' sites are grouped together and are less similar to any of the other sites before spraying in 1988, but in 1989 there is some sign that the now sprayed '88' sites have become less similar to other sites and more similar to '87' sites. Thus DDT may have some influence on species composition of sites, although sampling date has a greater effect on faunal similarity of sites than does the insecticide.

Relative abundance of individual taxa   Relative abundance of the 95 most numerous taxa caught was compared between sprayed and unsprayed areas and within the spray monitoring area over the two years of the study. There were significant differences detected between sprayed and unsprayed areas for many taxa, but few showed a consistent change in relative abundance which could be attributed directly to DDT. The mites and springtails did show some signs of effects (Box 3.7), but temporal and other uncontrolled environmental factors were the greatest cause of variation.

DDT in the Tropics

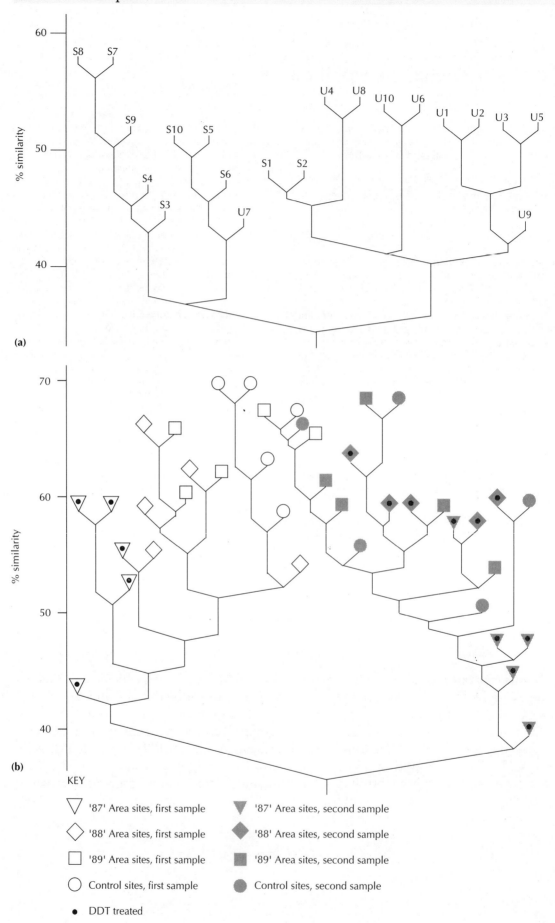

Figure 3.10 Hierachical classification of sites based on Sorenson's Quotient of Similarity. Whole Study Area (a) Sprayed (S) versus unsprayed (U) sites. SMA (b) Pre- and post-spray 1988 (c) Pre- and post-spray 1989 (d) Pre-spray 1988 and pre-spray 1989

Terrestrial Invertebrates

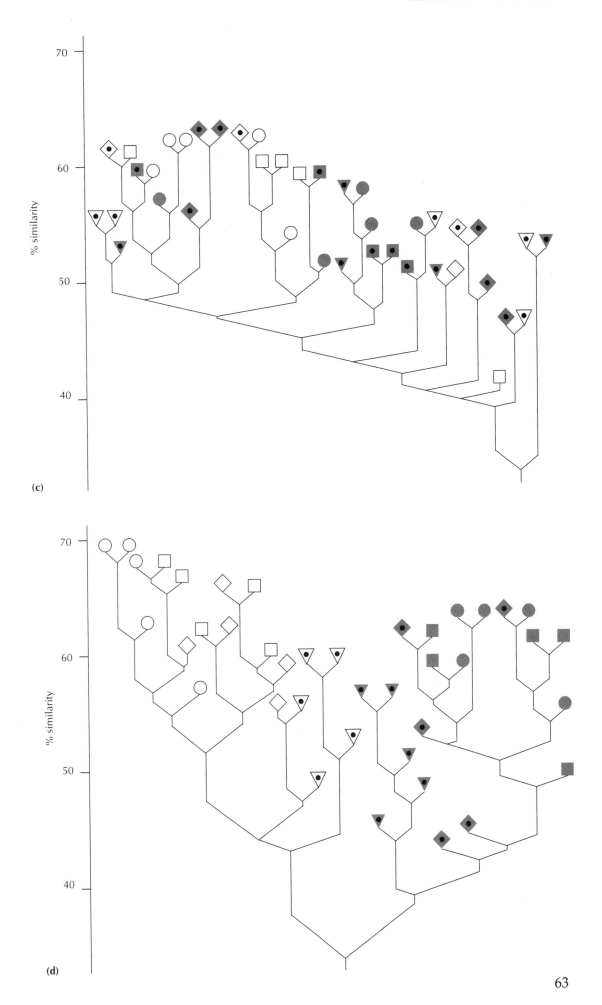

(c)

(d)

63

Box 3.7 Micro-invertebrate abundance

Few of the micro-invertebrates sampled show consistent evidence of an adverse impact of DDT on their relative abundance. Surface-active Cryptostigmata were more abundant in the sprayed (S) area than in the unsprayed (U), but trap catches in the spray monitoring area showed no evidence that this difference could be attributed to DDT, as fewer Crytostigmata were caught in the most heavily sprayed '87' area than in any of the other sectors, throughout the study (Figure 3.11). Thus no clear pattern is apparent for epigeal Cryptostigmata. However, a single species of soil-dwelling Cryptostigmata, *Hypozetes* sp. [Ceratozetidae] (Plate 14), occurred in significantly higher numbers in litter bags from repeatedly sprayed sectors of the SMA than from sectors sprayed once or not at all.[23] None of the other 37 species showed any differences in relative abundance between sprayed and unsprayed sectors.

Epigeal Mesostigmata were caught in small numbers throughout the study and there were too few to show statistically significant differences between sites. However, the data indicate that fewer Mesostigmata were caught in pitfall traps at sprayed sites (Figure 3.11), which is consistent with previous findings on DDT impacts.[7,40] This trend is confirmed by results for soil-dwelling Mesostigmata recovered from litter bags, with *Protogamasellus* sp. *mica* species-group [Ascidae] occuring in smaller numbers in sprayed sectors of the SMA.[23]

The epigeal Prostigmata showed enormous variability in numbers caught and few taxa showed consistent effects in response to spraying. The predatory Bdellidae, which have been shown to be very sensitive to DDT in temperate agricultural soils[41], showed no evidence of a decline in abun-

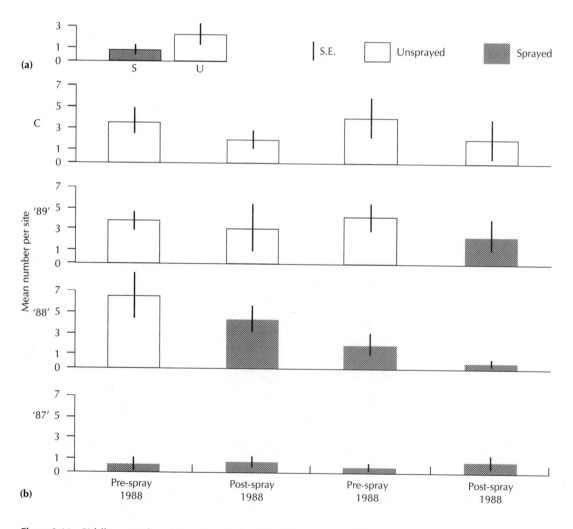

Figure 3.11 Pitfall trap catches of Mesostigmata (Acarina) in the general study area (U and S sites) and spray monitoring area over two years of the study

Terrestrial Invertebrates

dance following several DDT treatments in this study. However, the Nanorchestidae were consistently less abundant in sprayed areas and were completely absent from pitfall traps in the sprayed area. The habits of these small mites are not known. Soil-dwelling Prostigmata were extracted from litter bags in relatively small numbers and none showed changes in relative abundance attributable to DDT spraying. However, species richness of the Prostigmata was lower in the sprayed sectors of the SMA than in the unsprayed sector (Table 3.4; Figure 3.4).

Collembola showed no consistent differences in abundance between sprayed and unsprayed areas.

More Isotomidae were recorded in the sprayed (S) area than the unsprayed (U) area, which would be expected if DDT were affecting predators, as occurred in response to agricultural use of the insecticide. However, results from the SMA did not support any evidence of an effect, either from pitfall trap catches, trunk trap catches, soil cores or litter bag samples.

The Collembola and other micro-invertebrates were less abundant in the '87' sector of the spray monitoring area, supporting evidence that the fauna in this sector was impoverished, though not necessarily as a result of DDT use.

Several taxa were not caught at all in sprayed area pitfall traps, but were found in the unsprayed areas, including a previously undescribed woodlouse *Aphiloscia* sp. [Isopoda: Philosciidae]; and the ponerine ant, *Ophthalmopane berthoudi*. There was insufficient data to attribute this to the use of DDT, but these taxa may have been affected. Extensive surveying within the study area also uncovered only one individual *O. berthoudi* within the sprayed area and this was not in a tract of woodland which had been sprayed. This ant is often predatory on termites and may be susceptible to accumulation of DDT residues from its food. Another ponerine ant, *Platythyrea cribrinodis*, which was shown to accumulate high residue levels in sprayed areas (Box 3.8), also showed indications of population decline coincident with DDT spraying (Plate 15). It became increasingly scarce in the SMA and the sprayed area as a whole. There is no proof that this observed decline was due to DDT, but circumstantial evidence points strongly in that direction.

Pitfall trap catches of several other species of ant also showed significant differences between sprayed and unsprayed areas, although few were consistent in relation to DDT treatment. Pitfall trapping can give misleading results in attempting to estimate populations of ants,[43,44] so other sampling techniques, including nest surveys and the use of food baits, were also employed. An extensive survey of nests of the common species showed no differences in numbers or distribution around trees in the sprayed (S) and unsprayed (U) areas. Similarly, the use of food baits in 1990 failed to find evidence of effects of DDT on the abundance and species-richness of the ground-foraging ant community in the SMA. However, *P. cribrinodis* was absent from the catches, irrespective of sector, although it had been caught in pitfall traps there in 1988 and 1989. *P. cribrinodis* was attracted to similar baits elsewhere and its absence from the SMA in 1990 is further evidence of a DDT-associated decline.[42]

Arboreal arthropods

No consistent adverse effects of DDT on arboreal arthropods were apparent from the trunk trap catches. Only the Collembola showed significant differences in abundance between sprayed and unsprayed sectors of the SMA, with more caught in the sprayed sectors. However, the dominant family, the Entomobryidae, showed no differences.

The Diptera and the Hemiptera showed short-term changes in abundance on trees following spraying, but as individual families within these orders showed varying patterns of change, it is unclear whether there was a real effect of DDT. Individual families and species differ biologically and with classification at order level, there is insufficient evidence to attribute differences in relative abundance to an impact of DDT. Data were too sparse to allow analysis of differences in numbers of individual families or species for these groups.

Soil fauna

Only nematodes occurred in large enough numbers in soil cores to warrant statistical analysis and there was no evidence that DDT affected their relative abundance (Table 3.3). However, fewer species were extracted from litter bags in the sprayed sectors than in unsprayed (Table 3.4). This is consistent with previous findings of the effects of DDT on nematodes.

Table 3.3 Mean number of invertebrates from soil cores taken in sprayed and unsprayed areas.

Taxon	Unsprayed		Sprayed	
	$\bar{X}$	s.e.	$\bar{X}$	s.e.
Nematoda	167.00	23.22	164.00	20.58
Acarina	4.4	0.79	6.13	2.33
Collembola	2.67	1.37	1.27	0.32
Coleoptera	0.8	0.32	1.33	0.53
larvae	0.4	0.13	0.53	0.25
Diptera larvae	1.87	0.43	1.4	0.61
Isoptera	0.33	0.22	1.2	0.79
Psocoptera	0.2	0.1	0.13	0.13
Homoptera	0.2	0.14	0.07	0.06
Formicidae	2.6	2.1	10.9	8.75

Figure 3.12 There was no evidence of an adverse effect on termites themselves, although they were an important link in the uptake of residues in the food chain (*Source*: R.A. Johnson)

There was also evidence of effects of DDT on soil mites. Fewer species of Prostigmata were extracted from litter bags from the sprayed sectors than from the unsprayed sector of the SMA (Table 3.4).

Similar numbers of Astigmata, Cryptostigmata and Mesostigmata were found in the sprayed sectors of the SMA and the unsprayed and species richness was also similar. However, *Hypozetes* sp. [Cryptostigmata: Ceratozetidae] occurred in larger numbers and *Protogamasellus* sp. [Mesostigmata: Ascidae] in smaller numbers in sprayed sectors than in the unsprayed (Box 3.7).

Very few termites were found in soil cores or in litter bags and thus wooden baits were used in order to assess relative abundance and activity of these important insects in the woodland environment. Termites were more active around baits in sprayed woodland than unsprayed and in some cases attacked more baits around sprayed trees.[22] The only termite species found on baits when they were collected, were workers and soldiers of *Ancistrotermes latinotus* (Holmgren). However, it is possible that other genera and species may also have been involved in attacks on other baits. *A. latinotus* is the most common termite found in Zimbabwe and favours mopane woodland (amongst other habitats), feeding on a variety of woody material, leaf litter and living vegetation (Figure 3.12).[30]

Table 3.4 Mean number of species of nematodes, mites and insects extracted from litter bags buried in sprayed and unsprayed woodland

Taxon	'87' Area $\bar{X}$	s.e.	'88' Area $\bar{X}$	s.e.	'89' Area $\bar{X}$	s.e.	Control $\bar{X}$	s.e.
Nematoda	3.2	1.6	3.0	1.4	7.2	1.4	6.0	1.4
Acari	29.0	5.4	24.6	4.3	31.0	3.6	29.6	2.6
Astigmata	4.4	0.8	4.4	0.8	4.2	0.4	3.6	0.4
Cryptostigmata	14.2	2.6	13.0	2.3	12.4	1.6	15.4	2.0
Mesostigmata	5.4	0.9	2.8	0.6	5.8	1.1	3.8	0.4
Prostigmata	5.0	1.3	4.4	0.8	8.6	1.5	6.8	0.7
Insecta	3.6	1.4	2.8	1.2	4.4	1.4	6.8	0.9
Total	38.6	6.8	32.4	6.8	45.4	5.7	45.8	4.3

$\bar{X}$ = mean number per tree; 5 trees per area

The results contrast with those from experiments investigating the effects of DDT treatment to cleared forest soil in Nigeria, where fewer softwood baits were attacked by *Ancistrotermes* sp. and *Pseuacanthotermes* sp. in treated plots.[40]

INVERTEBRATES IN THE FOOD CHAIN

Insects and other invertebrates play an important part in the ecology of many higher animals and are the major food source for a wide range of reptiles, amphibians, birds and mammals. The contamination of invertebrates with DDT residues therefore has implications for the status of other animals in the ecosystem. These were considered through studies of diet in lizards (Box 4.1) and birds (Box 5.4) and by measurement of residue concentrations in samples of prey species (Box 3.8).

Residue analysis revealed that all the invertebrates sampled in the sprayed area contained DDT (Table 3.5). Many showed traces of the insecticide, even when taken in unsprayed areas, but residue burdens in insects taken from sprayed sites were always greater and in some cases reached very high levels. Ants showed the highest residue burdens of the fauna sampled, with most of the insecticide being in the form of p,p'-DDT, the most acutely toxic form. Termites and beetles on the other hand had generally lower DDT burdens and a higher proportion of this appeared as the metabolite DDE, which is less acutely toxic to invertebrates. This may have been a result of sampling, as ants were taken directly from sprayed tree trunks, whilst termites and beetles were generally collected from the ground in the immediate vicinity of the same trees. Despite the high residue burdens, none of the ants sampled showed any unco-ordinated behaviour or tremors, characteristic of DDT poisoning.

Box 3.8 DDT residues in invertebrates

Individual invertebrates were collected alive from sample sites, using clean forceps, and placed in small aluminium canisters or glass sample vials with screw tops lined with aluminium foil. Samples were either frozen or kept in 10% formalin solution, from time of collection until residue analyses were carried out. Forceps were washed in xylene and rinsed in alcohol before changing sample insect or sampling site, but not between individuals of the same species taken at the same sampling site.

In every case, except that of one cricket, the DDT burdens were considerably higher in samples from sprayed areas than from unsprayed. DDT residues from invertebrates in the unsprayed area (except for the one cricket mentioned above) were mainly in the form of p,p'-DDE, whereas invertebrates from the sprayed areas contained a high proportion of unchanged DDT (mainly the p,p'-DDT isomer).

The ant species *Camponotus* spp. and *Platythyrea cribrinodis* showed the highest residue levels. This parallels the findings of a survey of carnivorous insects in forest and agricultural areas in Belgium.[16] It was also notable that the highest proportion of DDT detected in these insects was in the form of p,p'-DDT, whereas the termites and Adesmini beetles collected had similar proportions of DDT and DDE, or a higher proportion of DDE. DDE is recognized as one of the first metabolites of DDT produced by a variety of invertebrates, and is much less acutely toxic to insects than DDT itself. The high level of unchanged DDT in the ants suggests that the species collected are tolerant to the insecticide without metabolizing it. Indeed one *P. cribrinodis* contained 2.9 µg p,p'-DDT and was still apparently healthy, showing no lack of co-ordination or tremors characteristic of DDT poisoning. This fact could be significant for predators feeding on these ants (Plate 16).

Table 3.5 Geometric mean residue levels (ppm dry weight) of DDT and its metabolites in bulked samples of ants, termites and beetles 6 months after the last spray treatment.

	N	N_i	DDE	DDD	DDT	ΣDDT	Sample range
Isoptera							
Unsprayed	6	539	0.40	0.00	0.33	0.69	0.18–2.38
Sprayed	12	1943	1.81	0.18	1.46	3.32	0.98–13.6
Camponotus sp.							
Unsprayed	6	258	0.45	0.00	0.00	0.45	n.d.–1.67
Sprayed	12	567	1.91	0.32	5.83	8.71	0.52–218.0
Coleoptera (Tenebrionidae:Adesmini)							
Unsprayed	5	20	0.36	0.00	0.30	0.79	0.32–2.23
Sprayed	11	51	0.65	0.05	0.34	0.92	0.14–8.12

N = number of samples
N_i = number of insects
n.d. = residues not detectable

DDT can persist in the biotic environment for long periods of time as evidenced by the 3.3 ppm ΣDDT recorded in *Platythyrea cribrinodis*, 2 years after the area in which it was caught had last been sprayed. Although levels were considerably lower than in individuals collected within 6 months of spraying, they were still much higher than in unsprayed areas.

Discussion

In general, natural spatial and temporal variation in relative abundance of invertebrates was high within the study area and any impact of DDT would have to be fairly severe to show up against this 'background noise'. The differences in abundance of soil-dwelling mites parallel previous findings and indicate that despite discriminative application, sensitive fauna do suffer direct effects of the DDT. No such major impact was detected in terms of relative abundance of any epigeal taxon, nor changes in diversity or community structure of the surface-active assemblages sampled. However, there were indications of both primary and secondary effects of DDT both at the community and individual level.

The mopane woodland ecosystem is a poorly studied one and the ecology of this habitat remains largely unknown. This is reflected in the large number of previously undescribed species recorded from pitfall traps[21]. There is little information on the biology and habits of many of the invertebrates found, even those to which names can be assigned. As a result, it is difficult to do more than guess at the ecological implications of any changes in the invertebrate fauna of the mopane woodland occurring as a result of DDT use. The results suggest that DDT-induced change in invertebrate abundance, if it occurs at all, is secondary to natural fluctuation in populations and activity. This is caused by microhabitat differences within different tracts of the mopane habitat and temporal and seasonal changes in temperature, relative humidity and other environmental factors. Nonetheless, evidence that DDT is having an influence on the invertebrate faunal composition within the woodland cannot be dismissed as insignificant. Until the ecology of this type of woodland is understood, any change in the fauna must be regarded as potentially leading to change in the functioning of the system.

Despite the fact that there is little evidence of population decline in the invertebrate groups sampled, their ability to carry high residue levels of DDT in their bodies and for it to persist over relatively long periods of time indicates that invertebrates are important components of ecological effects of DDT ground-spraying. The food chain link from invertebrates, to woodland birds (Chapter 5) and the skink *Mabuya striata* (Chapter 4) which suffered population decline in sprayed areas is clear.

References

1. Brown, A.W.A. (1978) *Ecology of Pesticides*. New York: John Wiley and Sons.

2. COPR (1978) *Pesticide residues research project*. Final report and recommendations 1975–1978. Ibadan, Nigeria (ODM research scheme R2730): Ministry of Overseas Development/International Institute of Tropical Agriculture.

3. Spain, A.V. (1974) The effects of carbaryl and DDT on the litter fauna of a Corsican pine (*Pinus nigra* var. *maritima*) forest: A multivariate comparison. *Journal of Applied Ecology*, **11**: 467–481.

4. Dindal, D.L., Folts, D. and Norton, R.A. (1975) Effects of DDT on community structure of soil microarthropods in an old field. In: *Progress in Soil Zoology*. (Vanek, J., ed.) pp. 505–513. The Hague: Junk Publishers.

5. French, N., Lichtenstein, E.P. and Thorne, G. (1959) Effects of some chlorinated hydrocarbon insecticides on nematode populations in soils. *Journal of Economic Entomology*, **52**: 861–865.

6. Sheals, J.G. (1956) Soil population studies. 1. The effects of cultivation and treatment with insecticides. *Bulletin of Entomological Research*, **47**: 803–822.

7. Edwards, C.A., Dennis, E.B. and Empson, D.W. (1967) Pesticides and the soil fauna: Effects of aldrin and DDT in an arable field. *Annals of Applied Biology*, **60**: 11–22.

8. Dempster, J.F. (1967) A study of the effects of DDT applications against *Pieris rapae* on the crop fauna. *Procceedings of the 4th British Insecticides and Fungicides Conference*, **1**: 19–25.

9. Butcher, J.W. and Snider, R.M. (1975) The effect of DDT on the life history of *Folsomia candida* (Collembola:Isotomidae). *Pedobiologia*, **15**: 53–59.

10. Hoffmann, C.H., Townes, H.K., Swift, H.H. and Sailer, R.I. (1949) Field studies on the effects of airplane applications of DDT on forest invertebrates. *Ecological Monographs*, **19**: 1–46.

11. DeBach, P. (1974) *Biological Control by Natural Enemies*. London: Cambridge University Press.

12. Davis, B.N.K. (1968) The soil macrofauna and organochlorine insecticide residues at twelve agricultural sites near Huntingdon. *Annals of Applied Biology*, **61**: 29–45.

13. Gish, C.D. (1970) Pesticides in soil. Organochlorine insecticide residues in soils and soil invertebrates from agricultural lands. *Pesticides Monitoring Journal*, **3**(4): 241–252.

14. Edwards, C.A. (1973) *Persistent Pesticides in the Environment*. Ohio: CRC Press.

15. Barker, R.J. (1958) Notes on some ecological effects of DDT sprayed on elms. *Journal of Wildlife Management*, **22**(3): 269–274.

16. Thomé, J.P., Debouge, M.H. and Louvet, M. (1987) Carnivorous insects as bioindicators of environmental contamination: Organochlorine insecticide residues related to insect distribution in terrestrial ecosystems. *International Journal of Environmental and Analytical Chemistry*, **30**: 219–232.

17. Koeman, J.H., Balk, F. and Takken, W. (1980) *The Environmental Impact of Tsetse Control Operations. A Report on Present Knowledge*. FAO Animal Production and Health Paper 7 rev.1. Rome: FAO.

18. Koeman, J.H. (1977) Effects of tsetse fly control measures on non-target organisms. *Mededelingen van de Faculteit Landbouwwetenschappen Rijksunivsiteit, Gent*, **42**(2): 889–896.

19. Tibayrenc, R. and Gruvel, J. (1977) La campagne de lutte contre les glossines dans le basin du lac Tchad. II Contrôle de l'assainissement glossinaire. Critique technique et financière de l'ensemble de la campagne. Conclusion générals. *Revue d'élevage et de médecine vétérinaire des pays tropicaux*, **30**(1): 31–39.

20. Pollock, J.N. (1982) *Training Manual for Tsetse Control Personnel*. Vol. II *Ecology and Behaviour of Tsetse*. Rome: FAO.

21. Tingle, C.C.D., Lauer, S. and Armstrong, G. (1992) Dry season, epigeal invertebrate fauna of mopane woodland in northwestern Zimbabwe. *Journal of Arid Environments*, **23**: 397–414.

22. Douthwaite, R.J. and Tingle, C.C.D. (1992) Effects of DDT treatments applied for tsetse fly control on White-headed Black Chat (*Thamnolaea arnoti*) populations in Zimbabwe. Part II: Cause of decline. *Ecotoxicology*, **1**: 101–115.

23. Tingle, C.C.D. and Grant, I.F. (in press) The effect of DDT on litter decomposition and soil fauna in semi-arid woodland. *Acta Zollogica Fennica*.

24. Grant, I.F. (1989) Monitoring insecticide side-effects in large-scale treatment programmes: Tsetse spraying in Africa. In: *Pesticides and Non-target Invertebrates*. (Jepson, P.C., ed.) Wimborne, Dorset, U.K: Intercept, pp. 43–69.

25. Disney, R.H.L. (1990) Revision of the Alamirinae (Diptera: Phoridae). *Systematic Entomology*, **15**: 305–320.

26. Disney, R.H.L. (1991) Scuttle flies from Zimbabwe (Diptera: Phoridae) with the description of five new species. *Journal of African Zoology*, **105**: 28–42.

27. Disney, R.H.L. (1992) The 'missing' males of the Thaumatoxeninae (Diptera: Phoridae). *Systematic Entomology*, **17**: 55–58.

28. MacFarlane, D. (Personal Communication), International Institute of Entomology, c/o Department of Zoology, The Natural History Museum, Cromwell Road, London SW7 5BD, UK.

29. Murphy, J.A. (Personal Communication), 323 Hansworth Road, Hampton, Middlesex.

30. Mitchell, B.L. (1980) Report on a survey of the termites of Zimbabwe. *Occasional Papers. National Museum Rhodesia Ser. B, Natural Science*, **6**(5): 187:323.

31. Moeed, A. and Meads, M.J. (1983) Invertebrate fauna of 4 tree species in the Orongorongo Valley, New Zealand, as revealed by trunk traps. *New Zealand Journal of Ecology*, **6**: 39–53.

32. Wallwork, J.A. (1976) *The Distribution and Diversity of Soil Fauna*. London: Academic Press.

33. Raw, F. (1959) Estimating earthworm populations by using formalin. *Nature*, **184**: 1661–1662.

34. Lavelle, P. (1988) Assessing the abundance and role of invertebrate communities in tropical soils: Aims and methods. In: Proceedings of a seminar on resources of soil fauna in Egypt and Africa. Cairo, 16–17 April 1986. (Bhabbour, S.I. and Davis, R.C., eds) *Journal of African Zoology*, **102**: 275–283.

35. Lasebikan, B.A. (1988) Studies in soil fauna in Africa: current status and prospects. In: Proceedings of a seminar on resources of soil fauna in Egypt and Africa. Cairo, 16–17 April 1986. (Ghabbour, S.I. and Davis, R.C., eds.) *Journal of African Zoology*, **102**: 301–311.

36. Block, W. (1970) Micro-arthropods in some Uganda soils. In: *Methods of Study in Soil Ecology*. (Phillipson, J., ed.) Proceedings of the Paris symposium organized by UNESCO and the International Biological programme. pp 195–202. Paris: UNESCO.

37. Rogers, L.E., Hinds, W.T. and Buschbom, R.L. (1976) A general weight vs. length relationship for insects. *Annals of the Entomological Society of America*, **69**: 387–389.

38. Rogers, L.E., Buschbom, R.L. and Watson, C.R. (1977) Length-weight relationships of shrub-steppe invertebrates. *Annals of the Entomological Society of America*, **70**: 51–53.

39. Tingle, C.C.D. (unpublished) The effects of DDT on diversity and faunal composition of invertebrates in mopane woodland in northwestern Zimbabwe.

40. Perfect, T.J., Cook, A.G., Critchley, B.R. and Russell-Smith, A. (1981) The effect of crop protection with DDT on the micro-arthropod population of a cultivated forest soil in the sub-humid tropics. *Pedobiologia*, **21**: 7–18.

41. Wallace, M.M.H. (1954) The effect of DDT and BHC on the population of the lucerne flea, *Sminthurus viridis* (L.) (Collembola) and its control by predatory mites, *Biscirus* spp. (Bdellidae). *Australian Journal of Agricultural Research*, **5**: 148–155.

42. Tingle, C.C.D. (1993) Bait location by ground-foraging ants (Hymenoptera: Formicidae) in mopane woodland selectively sprayed to control tsetse fly (Diptera: Glossinidae) in Zimbabwe. *Bulletin of Entomological Research*, **83**: 259–265.

43. Douthwaite, R.J., Fox, P.J., Matthiessen, P. and Russell-Smith, A. (1981) *The Environmental Impact of Aerosols of Endosulfan Applied for Tsetse Fly Control in the Okavango Delta*, Botswana. Final report of the endosulfan monitoring project, London: Overseas Development Administration.

44. Marsh, A.C. (1984) The efficacy of pitfall traps for determining the structure of a desert ant community. *Journal of the Entomological Society of South Africa*, **47**(1): 115–120.

Plates 1–4 Tsetse Fly control

Tsetse flies feed on blood and transmit sleeping sickness to Man and his livestock. Twenty-two species occur in Africa, infesting 11 million km². (Plate 1: Source, CDE, Porton Down).

Three tsetse control techniques have been used in recent years. Ground-spraying with persistent pesticides (Plate 2: R.J. Douthwaite), such as DDT, has been the most successful; aerial spraying (Plate 3: R. Allsop) with non-persistent doses of endosulfan has few side-effects and can work well in flat country; the effectiveness of tsetse-attractive, odour-baited, insecticide-impregnated cloth 'targets' (Plate 4: R.J. Douthwaite) remains in doubt despite their current popularity.

Plates 5–8   The Spraying Operation

Ground-spraying is labour-intensive (Plate 5: H.Q.P. Crick) and requires good access, the construction of seasonal camps and sturdy vehicles (Plate 6: I.F. Grant). The insecticide is applied from knapsack compression sprayers (Plate 7: H.P.Q. Crick) to tsetse fly resting sites in woodland and along streams (Plate 8: H.P.Q. Crick).

Plates 9–12 Habitats at Siabuwa

Streams near Siabuwa are seasonal and subject to flash floods during the rainy season (Plate 9: R.J. Douthwaite, Plate 10: B. McCarton). The woodland is deciduous (Plate 11: R.J. Douthwaite) and often scrubby due to shallow soils, fire and widespread elephant damage (Plate 12: R.J. Douthwaite).

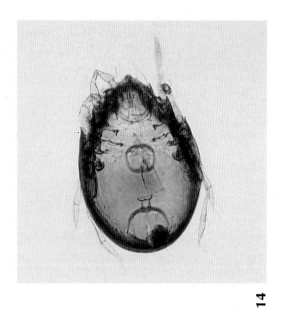

13

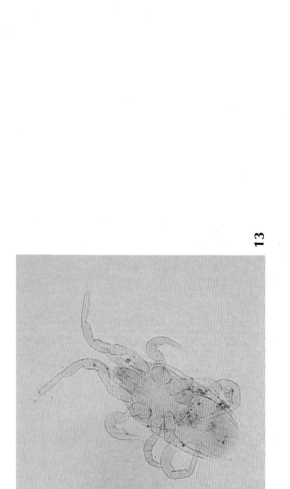

15

14

16

Plates 13–16 Soil Invertebrates

Heavy DDT applications on farmland reduce the number of predatory mites in soil, but as predation declines the abundance of some detrivorous mites increases. In this study, the relative abundance of *Protogamasellus* sp. (Mesostigmata: Ascidae) (Plate 13: C.C.D. Tingle), a predator, and *Hypozetes* sp. (Cryptostigmata: Ceratozetidae) (Plate 14: C.C.D. Tingle), a detritivore, in sprayed and unsprayed areas, were consistent with these effects of DDT.

Predatory ants fed on termites and other insects and were heavily contaminated with DDT. *Platythyrea cribrinodis* (Plate 15: C.C.D. Tingle) disappeared from sprayed plots after treatment, while *Ophthalmopane berthoudi* was absent from the sprayed area yet common and widespread in the unsprayed area. *Camponotus* spp. ants (Plate 16: Jane Burton, Bruce Coleman), however, were common in the sprayed area passing their residues to White-headed Black Chats and predators higher in the food chain.

Plates 17–20 Tree Trunk Fauna

Animals using tree trunks, and especially holes, were at the greatest risk of contamination. Populations of the skink, *Mabuya striata* (Plate 17: C.C.D. Tingle), Red-billed Wood-hoopoe (Plate 18: Peter Johnson, NHPA), White-headed Black Chat (Plate 19: P. Steyn, Ardea) and the African Goshawk (Plate 20: P. Pickford, NHPA), which feeds on woodland birds and lizards, all declined in sprayed areas.

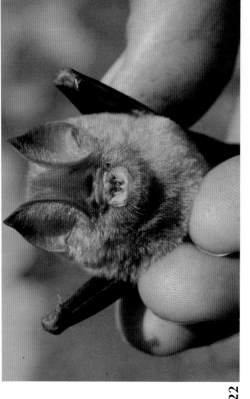

21

23

22

24

Plates 21–24 Bats and Bushbabies

Poor breeding success was recorded at a colony of *Rhinolophus hildebrandtii* (Plate 21: A.M. Hutson) roosting in a hollow tree which had been sprayed with DDT. Adult *Rhinolophus hildebrandtii*, *Hipposideros caffer* (Plate 22: A.M. Hutson) and *Nycteris thebaica* (Plate 23: A.M. Hutson) accumulated high concentrations of DDT residues, which may have proved fatal during food shortages.

Bushbabies may be exposed to high levels of DDT as they live in hollow trees, but numbers in sprayed and unsprayed areas were similar (Plate 24: C.C.D. Tingle).

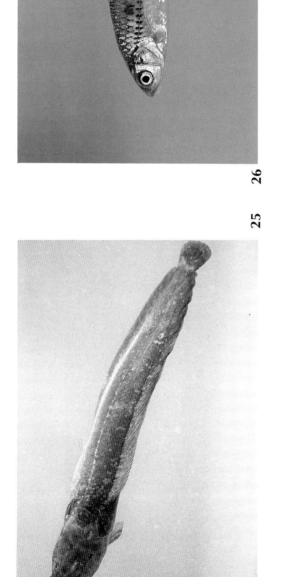
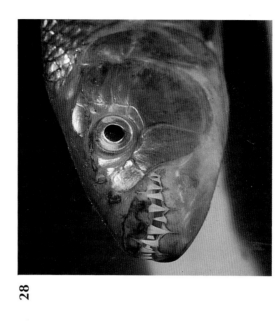

Plates 25–28 Fish

Sharptooth Catfish (Plate 25: G. Merron), Red-eye Labeo and Linespotted Barbs (Plate 26: P. Skelton) were the only common fish in seasonal streams near Siabuwa. High concentrations of DDT residues were found in samples from the sprayed section of the Songu river but effects were not confirmed.

On Lake Kariba, Butter Catfish from sprayed sites were significantly more contaminated than those at unsprayed sites, but growth and mortality appeared unaffected (Plate 27: Hans Reinhard, Bruce Coleman). Some Tigerfish accumulated high residue levels due to their predatory habits (Plate 28: G.D. Plage, Bruce Coleman).

29 **30**

31 **32**

Plates 29–32 Waterbirds

Fish Eagles (Plate 29: Nigel Dennis, NHPA) often fed on heavily contaminated fish such as Sharptooth Catfish and Tigerfish. Their eggs were collected from Lake Kariba by local climbers (Plate 30: R.J. Douthwaite) and from an airborne helicopter, fitted with a hoist. High concentrations of DDT residues in eggs at the eastern end of Kariba resulted in poor hatching success, but did not reduce the density of nesting eagles (Plate 31: Ardea).

Reed Cormorants also accumulated high concentrations of DDT residues and eggshell thinning and hatching failure were predicted (Plate 32: R.J. Douthwaite).

4

LIZARDS

M R K Lambert

Introduction

Lizards are abundant in the mid-Zambezi valley, forming an important link in the food chain between arthropods and birds of prey. They may be good indicators of environmental contamination for they have limited powers of dispersal and can be susceptible to DDT applied for tsetse fly control.[1] The impact of spraying operations in Zimbabwe was investigated during short visits in May 1989 and May–August 1990.

Surveys of relative abundance and age structure were made in mopane woodland and on rock outcrops at Siabuwa and in the Omay Communal Area and Matusadona National Park.

Nineteen species were recorded. Most are widespread in eastern and southern Africa. *Lygodactylus chobiensis*, although common, is restricted to the Zambezi valley, from the Okavango Delta in Botswana to Tete in Moçambique.[2]

There was no evidence that spraying reduced species richness. Seventeen species were found in woodland: one was absent from treated woodland and one from untreated woodland; both were scarce. Eleven species were found on outcrops: two were absent from sprayed sites and four from unsprayed sites; none was numerous.

Lizards in Woodland

A detailed study was made of numbers and DDT contamination in the skink *Mabuya striata*, the most abundant lizard and only one to be found at every sample site at Siabuwa (Plate 17).[3]

The frequency of occurrence of *M. striata* relative to other species declined significantly with increasing number of spray treatments (Analysis of Variance, $F_{1'15} = 5.8, P < 0.05$) (Figure 4.1). A similar trend was apparent comparing occurrence at sites in Matusadona (unsprayed) and Omay (treated twice).

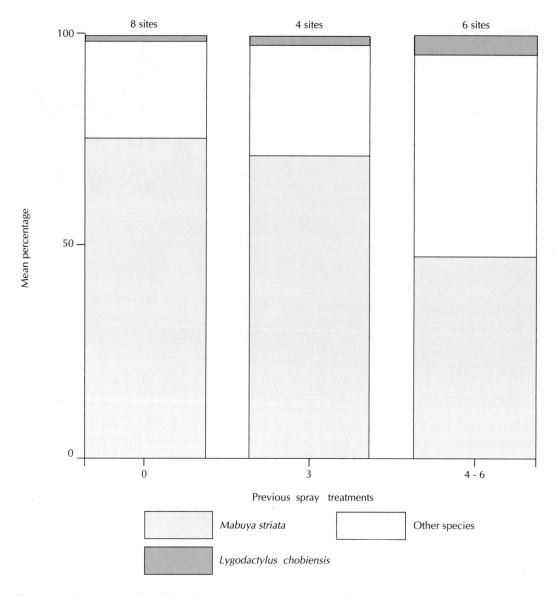

Figure 4.1 Composition of lizard fauna in mopane woodland at Siabuwa, by spray treatment

Actual abundance was also lower in sprayed areas (Table 4.1). Similarly, significantly more trees were occupied by *Mabuya striata* at unsprayed sites at Siabuwa than at sites sprayed 3–6 times. Numbers ranged from 6–15/300 trees ($n = 7$, mean 9.9) at unsprayed sites compared with 2–10/300 trees ($n = 7$, mean 6.3) at sprayed sites ($\chi^2 = 5.75$, $p < 0.02$).

Table 4.1 Sighting rates of lizards in sprayed and unsprayed woodland at Siabuwa (median and range seen per search-hour per site)

	Unsprayed ($n = 8$)	Sprayed ($n = 10$)	Mann-Whitney p
Mabuya striata	9.9 (6.1–19.3)	5.7 (1.3–12.2)	< 0.02
Other species	3.2 (0.5–5.7)	2.2 (0.4–9.6)	n.s.

Relatively more sub-adult *M. striata* were seen at sprayed sites, but age structure varied less between spray treatments than between study areas (i.e. Siabuwa and Omay/Matusadona). At Siabuwa an average of 91% of skinks were adult at unsprayed

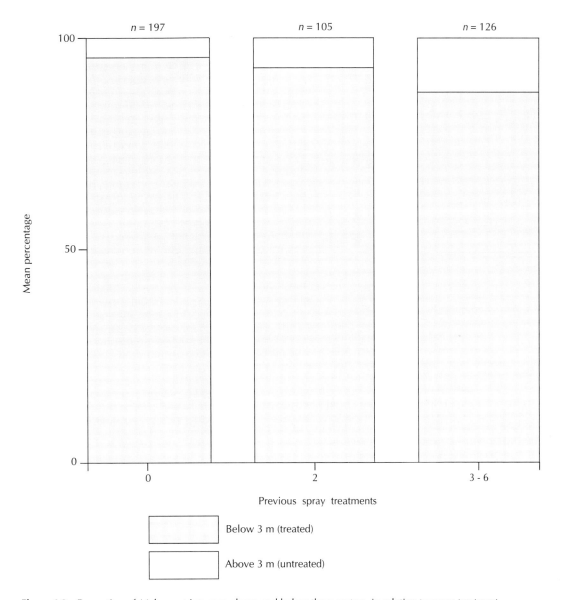

Figure 4.2 Proportion of *Mabuya striata* seen above and below three metres, in relation to spray treatment

sites and 84% at sprayed sites (3–6 x), compared with 65% at Matusadona (unsprayed) and 59% in Omay (sprayed twice).

Ninety-nine percent of *M. striata* ($n = 454$) were found on trees likely to be treated with DDT (i.e. > 15 cm diameter). Significantly more occurred above the spray target (i.e. above 3 m) in sprayed areas than in unsprayed areas ($X^2 = 8.9$, 2 d.f., $p < 0.02$) (Figure 4.2).

ΣDDT residue levels in *M. striata* increased significantly with exposure to annual treatments, taking account of the fact that juvenile and immature lizards in areas sprayed three or more times had not been exposed to more than one and two treatments respectively ($r_{34} = 0.67$, $p < 0.001$) (Table 4.2). DDE comprised 63% of the ΣDDT in lizards from unsprayed areas ($n=7$) and 56% in lizards from sprayed areas ($n=27$).

Table 4.2 ΣDDT levels in *Mabuya striata*, *M. quinquetaeniata* and *Agama kirkii* related to exposure to annual treatments.

	Treatments (n)	Sample size (n)	Geometric mean (range) ΣDDT ppm lipid
Woodland			
Mabuya striata	0	7	1.32 (0.21–6.51)
	1	2	6.58 (5.23–8.27)
	2	8	24.31 (2.65–64.35)
	3 or more	18	25.44 (2.84–263.34)
Outcrops			
Mabuya quinquetaeniata	0	6	0.31 (0–1.04)
	1 or more	4	1.22 (0.18–4.18)
Agama kirkii	0	1	0.24
	2 or more	4	0.54 (0–4.11)

LIZARDS ON ROCK OUTCROPS

Mabuya quinquetaeniata was the most abundant species at 12 outcrops, but *M. varia* was also recorded at every site. The frequency of occurrence of these species did not vary with spray treatment, but *Agama kirkii* was relatively more frequent in treated areas ($F_{1,19} = 29.6$, $p < 0.01$) (Figure 4.3).

The sighting rate of lizards did not vary between sprayed and unsprayed areas except for *Agama kirkii* (Table 4.3).

Table 4.3 Lizard numbers on sprayed and unsprayed outcrops at Siabuwa (median and range seen per search-hour per site).

	Unsprayed ($n = 4$)	Sprayed 5–6x ($n = 6$)	Mann-Whitney p
M. quinquetaeniata	13.9 (12.0–23.2)	12.3 (9.0–25.4)	n.s.
Mabuya varia	3.9 (2.6–8.0)	3.8 (3.0–4.4)	n.s.
Agama kirkii	0.7 (0–2.6)	4.0 (2.5–4.9)	<0.01
Other species	1.8 (0–3.5)	1.9 (0–2.0)	n.s.

Residue levels in *A. kirkii* and *M. quinquetaeniata* from outcrops showed little evidence of accumulation and were much lower than in *Mabuya striata* (Table 4.2).

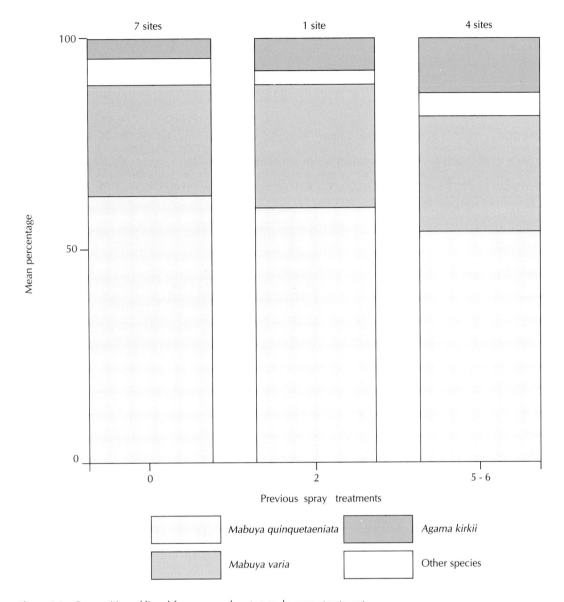

Figure 4.3 Composition of lizard fauna on rock outcrops, by spray treatment

DISCUSSION AND CONCLUSIONS

The results are consistent with reduced survival of *M. striata* after several years exposure to DDT. Fewer lizards were present in areas sprayed more than twice, and the proportion of mature lizards living on tree trunks below 3 m, the target zone for spray applications, was reduced. Residue levels increased with years of exposure and relatively high levels were found in lizards exposed to two or more spray applications. No evidence was found that other species accumulate high residue levels or are adversely affected; indeed, numbers of the dwarf gecko *Lygodactylis chobiensis*, which lives alongside *Mabuya striata*, increased in sprayed areas. The two species may compete for refuges, but they differ in diet.[2] *Lygodactylus* feeds mainly on subsurface termites and *Mabuya* on a wide range of invertebrates (Box 4.1). As a result, they are likely to accumulate different amounts of ΣDDT.

Box 4.1 Invertebrate prey for lizards
C.C.D. Tingle

A variety of invertebrates were found in the guts of the four species examined. The common tree-dwelling skink, *Mabuya striata*, favoured beetles and termites (Table 4.4), as did the other species of skink. The diet of *M. quinquetaeniata* (n = 10) was very similar, although a higher proportion had fed on Lepidoptera larvae; beetles were absent from the three *M. varia* guts examined. Various Orthoptera, Hemiptera and Hymenoptera were also commonly taken by all the skinks, although ants were only a minor food item. Ants were taken in larger numbers by *Agama kirkii* (n = 4) on rocky outcrop habitats and comprised the major food for these lizards.

The findings for *M. striata* are very similar to those made in the Kora National Reserve in Kenya on the same or closely related species.[4]

Table 4.4 Invertebrates in the guts of *Mabuya striata*

	Guts n	%	Prey n	items %
Total examined	44	100	1090	100
Diplopoda	1	2	1	+
Chilopoda	1	2	1	+
Araneae	5	11	5	+
Acarina: Caeculidae	1	2	1	+
Blattodea	1	2	1	+
Mantodea	1	2	0	+
Hemiptera	8	18	9	+
Homoptera	2	5	2	+
Fulgoroidea	2	5	2	+
Heteroptera	6	14	7	+
Reduviidae	4	9	4	+
Isoptera	25	57	951	87
Odontotermes sp.	11	25	641	59
Macrotermes sp.	1	2	45	4
Hodotermes mossambicus	8	18	80	7
Orthoptera	7	16	9	+
Acrididae: (indet.)	3	7	4	+
Acorypha sp.	1	2	1	+
Thericleidae	1	2	1	+
Lepidoptera: larva	2	5	2	+
Coleoptera	36	82	62	6
Progonochaetus incrassatus	3	7	5	+
Scarabaeoidea	1	2	1	+
Buprestidae	1	2	1	+
Tenebrionidae: *Zophosis* sp.	7	16	9	+
Chrysomelidae: Cassininae	1	2	1	+
Curculionidae	5	11	9	+
Hymenoptera	21	48	46	4
Formicidae	18	41	40	4
Pheidole spp.	7	16	23	2
Ophthalmopane berthoudi	4	9	6	+
Scolioidea: Mutillidae	2	5	4	+
Others	2	5	2	+

+ less than 1%

REFERENCES

1. Koeman, J.H., Den Boer, W.M.J., Feith, A.F., De Iongh, H.H. and Splietoff, P.C. (1978) Three years' observations on side effects of helicopter applications of insecticides used to exterminate *Glossina* species in Nigeria. *Environmental Pollution*, **15**: 31–59.

2. Branch, B. (1988) *A Field Guide to the Snakes and Other Reptiles of Southern Africa*. London: New Holland.

3. Lambert, M.R.K. (1993) Effects of DDT ground-spraying against tsetse flies on lizards in NW Zimbabwe. *Environmental Pollution*, **82**: 231–237.

4. Rotich, D., Hebrard, J. and Duff-Mackay, A. (1986) Analysis of gut contents of *Agama agama*, *Agama ruppelli*, *Mabuya quinquetaeniata* and *Mabuya maculilabris* from Kora National Reserve. pp. 241–244 In: *Kora: An ecological inventory of the Kora National Reserve, Kenya*. (Coe, M. and Collins, N.M., eds.) London: Royal Geographic Society.

5

WOODLAND BIRDS

R J Douthwaite

Introduction

Adverse effects on birds, and particularly effects of DDE on birds of prey, contributed significantly to the decision to ban DDT in many countries. Bird studies were therefore a major concern of this project.

DDT residues can kill, cause eggshell thinning and modify the behaviour of birds.[1,2] Songbird mortality provided the first evidence of adverse effects of DDT on wildlife forty-five years ago[3], yet there remains little information on the impact of exposure on songbird populations. Detailed monitoring schemes, such as the British Common Bird Census, were introduced to track the fate of exposed populations but were too late to measure the initial impact.[4] In Britain, DDT was rarely implicated in bird kills,[5,6] but in the United States of America, heavy mortality was associated with Dutch Elm disease control programmes, in which DDT was applied to elm trees *Ulmus americana* L. to control the beetle vector of the disease. More than 90 bird species were killed by these operations and population declines in some species reported[7] but these were confirmed only for the American Robin at a local level.[7–13] Henny[14] found no change in adult mortality rates in 16 species comparing the periods 1916–1945 and 1946–1965 and concluded that the reported declines in the American Robin were either very localized or were compensated for by a reduction in natural mortality.

Eggshell thinning was first linked with DDT residue levels in 1967[15] and has since been shown to affect hatching and breeding success in a wide range of flesh- and fish-eating birds, including pelicans, cormorants, darters, herons, ibises, ducks, vultures, hawks, eagles, falcons, gulls, owls and crows.[1] Depending upon species, DDE levels of 2–20 ppm wet weight in eggs (c. 13–130 ppm dry weight) are associated with breeding failure on a scale sufficient to cause population decline.[16–22] Irrespective of species, raptor populations declined when eggshell thinning exceeded 16–18% over several years[23] and, in some cases, were locally exterminated.

Over 635 species have been recorded in Zimbabwe[24] of which half (318) are known to occur around Siabuwa.[25] The list for the Zambezi valley includes cosmopolitan species and those endemic to local habitats, such as mopane and miombo (*Julbernardia*) woodlands, crags, riverine thicket, freshwater and their man-made derivatives; none is endemic to Zimbabwe and at risk of extinction from ground-spraying, although some, such as the Taita falcon, are local and rare and at risk of exposure to DDT residues.

Infrequent reports of songbirds dying in convulsions in areas recently sprayed for tsetse fly control are indicative of poisoning.[26–28] DDE levels exceeding 8 ppm wet weight have also been reported in eggs of nine out of 14 predatory species examined.[29] Species feeding on birds or fish were at greatest risk with highest residue levels in Peregrine and Lanner falcons, Black Sparrowhawk, African Goshawk, Fish Eagle, Wahlberg's Eagle, Pel's Fishing Owl, Darter and Grey Heron. Eggshell thinning in the Fish Eagle averaged 15.6% (J. Tannock[30]) and Thomson[29] predicted the species would be extinct on Lake Kariba within 10–15 years, and falcons and hawks in less, unless the use of DDT was curtailed. Monitoring of the Fish Eagle population breeding in Matusadona National Park, on Lake Kariba, began in 1975[31] and nest sites of species used in falconry have been recorded by the Zimbabwe Falconers' Club since 1976.[32] However, the impact of DDT on species populations has not been investigated either in Zimbabwe or elsewhere in the tropics.

Effects on Songbirds

Relative Abundance

Surveys of relative abundance of woodland songbirds at Siabuwa were made to identify differences between sprayed and unsprayed areas (Box 5.1).

Box 5.1 Counting Songbirds at Siabuwa
R.J. Douthwaite and H.Q.P. Crick

Surveys of relative abundance of songbirds were made in mixed mopane-*Combretum-Julbernardia* woodland in the northern and southern parts of the study area to identify population differences between sprayed and unsprayed areas.

In 1987, counts were made at 80 fixed points, 20 on each of two tracks passing through an area treated annually with DDT from 1984–86, and 20 on each of two tracks in an untreated area (Figure 5.1). Tracks were traversed by vehicle and sample points lay 0.4–0.5 km apart so that birds seen near one point were unlikely to be seen or heard at the next. Sampling began about 1 h after dawn, at c. 06.45 h, and continued for about 3 h. All points on one track were visited during one census. Five censuses were made along each track between 19 March and 8 April.

At each sample point the observer counted all birds seen or heard within 100 m of the centre point during 5 min. He was free to move to increase the probability of recording secretative or silent birds. If a mixed-species flock was encountered, up to 10 min was allowed to complete enumeration.

Further point count surveys were made between 4–18 May 1988 and 18–26 April 1989 on a reduced scale. In October 1989 birds were counted during a series of eight walks through sprayed and unsprayed areas. Each transect traced a square, avoided tracks, and lasted 40 minutes. All birds seen or heard within 50 m of the observer were counted within 5 min periods. This survey reduced any seasonal and spatial biases associated with the point count surveys made earlier.

Additional surveys, based on timed counts or response to playback of song on a tape recorder, were made of selected species to obtain more precise information.

A similar number of species was recorded during surveys in the sprayed north and unsprayed south of the study area. Ignoring overflying birds, and combining point counts from 1987 and 1989, 83 species were seen in the north and 87 species in the south; 69 were common to both areas.

Depending upon the survey, 35–39% fewer birds were seen in the north of the study area than in the south. Of the 33 most numerous species, six were commoner in the north and twenty-six in the south with one species equally abundant in both areas (Figure 5.1).

The climate was similar in the north and south, but the north was generally more hilly with sandier soils and less mopane woodland. In 1987 detailed observations were made of the vegetation to see whether differences in bird numbers were due to differences in vegetation.

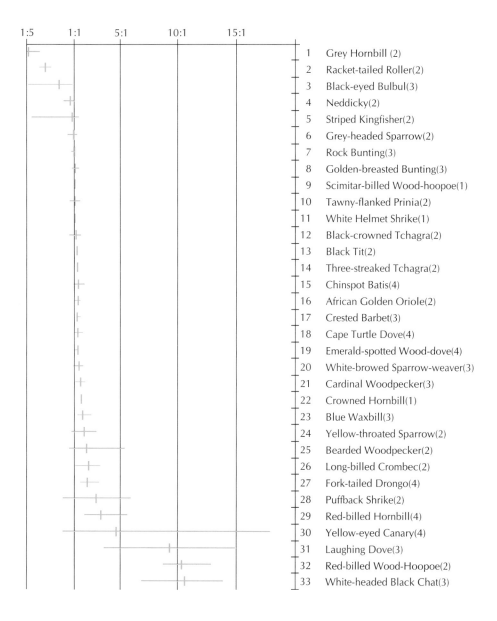

Figure 5.1 Relative abundance (mean and range) of 33 common bird species in the north and south of the study area (number of censuses in parentheses after species name)

Box 5.2 Vegetation Sampling
R.J. Douthwaite and H.Q.P. Crick

The woodland at each sample site was assigned to one of five types depending on the relative abundance of *Julbernardia* (J) and mopane (M), namely Type J: mopane absent; Type J>M: mopane scarce; Type J=M: *Julbernardia* and mopane both common; Type M>J: *Julbernardia* scarce; and Type M: *Julbernardia* absent. More detailed observations were then made on the vegetation at each point in two 0.01 ha plots. The densities of trees, 'small' shrubs (0.5–1 m high), 'medium' shrubs (1–2 m high), 'large' shrubs (2–3 m high) and 'very large' shrubs (>3 m high), as well as canopy cover (%) and ground cover (%) were estimated using modifications of the techniques of James and Shugart.[60]

It was found that apart from the prevalence of mopane woodland, southern sites were also characterized by significantly less ground cover, fewer small and medium shrubs, and more trees than northern sites. Large and very large shrub densities and canopy cover were similar. The comparative lack of ground cover in the south in 1987, when the observations were made, may have been partly due to drought conditions which affected the north less.

Fewer birds were seen in *Julbernardia* woodland in the north than in mopane, but there was little difference in numbers in the south, although few sites dominated purely by *Julbernardia* were sampled (Figure 5.2).

Using data from 1987, significant correlations were found between the vegetation at sample points and numbers of some of the 19 most abundant species (Table 5.1). These were used in comparing numbers in the north and south of the study area. It was found that 9 of the 19 species were significantly less common in the north, while 10 species were equally abundant (Table 5.2). The former group mainly comprised birds feeding at or near the ground on arthropods while the latter group were mainly species feeding in the tree canopy on seeds, fruit and insects.

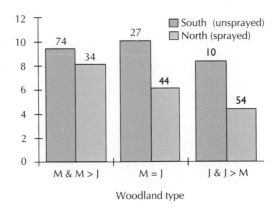

Figure 5.2 Average number of birds/site related to spray treatment and woodland type (all censuses included; White-browed Sparrow Weaver excluded; sample size at head of histogram)

Many factors, including climate, food supply, availability of nesting and roosting sites, and predation can affect bird numbers. Climatic differences and differences in eight measures of woodland type and woodland structure do not explain the relative scarcity of almost half the species sampled in the north of the study area. However, other environmental differences may have been important. The choice of vegetative parameters was arbitrary and there is no indication that the measures chosen, the scale of measurement, or the number of samples taken were appropriate for every bird species. Indeed, given the varied requirements of birds, it would be remarkable if they were. For example, no direct account was taken of the availability of large seeds, which are important for doves, or of harvester termite *Hodotermes mossambicus* workers, which are eaten in large amounts by Red-billed Hornbills, and it is possible that the association of these key factors with the proxy measures chosen (e.g. mopane woodland, with few shrubs and low ground cover) were too weak to be correlated significantly. Unquantified natural factors, perhaps linked to the sandier soils and drier conditions under *Julbernardia* woodland, could therefore explain the comparative scarcity of many species in the north. However, a short-term impact of DDT on macro-arthropods in sandy soil was detected (Chapter 2) and the possibility that this affected bird numbers in *Julbernardia* woodland, directly or indirectly, cannot be discounted.

Table 5.1 Vegetation correlates* with bird numbers in 1987

Species	Woodland type	Woodland structure
Grey Hornbill	None	More ground cover, more hornbills
Black-eyed Bulbul	Few in J	More medium shrubs, more bulbuls
Scimitar-billed Wood-hoopoe	More in J<M	None
Black Tit	None	None
African Golden Oriole	Few in J	More small shrubs, more medium shrubs, less ground cover, fewer orioles
White-browed Sparrow-weaver	Colonies only in M, M>J and M=J	None
Cape Turtle Dove	None	More ground cover, more doves
Emerald-spotted Wood Dove	None	None
Cardinal Woodpecker	None	None
Crowned Hornbill	None	More trees, more hornbills; More medium shrubs, fewer hornbills
Crested Barbet	More in J<M	More ground cover, more barbets
Chinspot Batis	None	None
Long-billed Crombec	None	More small shrubs, fewer crombecs
Yellow-eyed Canary	None	None
Fork-tailed Drongo	None	None
Red-billed Wood-hoopoe	None	More ground cover, fewer wood-hoopoes
Red-billed Hornbill	Few in J	None
Bearded Woodpecker	None	More ground cover, more woodpeckers
White-headed Black Chat	None	None

* Spearman's Rank Correlation, $P < 0.05$

Table 5.2 Ratios of relative abundance of some of the commoner species in the north and south of the study area, related to feeding habits.

	Relative Abundance		Feeding Station	Diet
	Ratio*	Significance†		
Grey Hornbill	−4.9	n.s.	Canopy, branches	Animals
Black-eyed Bulbul	−2.3	n.s.	All levels	Fruit, insects
Scimitar-billed Wood-hoopoe	1.1	n.s.	Branches	Arthropods
Black Tit	1.3	n.s.	Branches	Insects
Chinspot Batis	1.4	n.s.	Canopy	Insects
African Golden Oriole	1.4	n.s.	Canopy	Insects, fruit
Crested Barbet	1.4	n.s.	All levels	Insects, fruit
Cape Turtle Dove	1.4	***	Ground	Seeds, insects
Emerald-spotted Wood-dove	1.4	**	Ground	Seeds, insects
White-browed Sparrow-weaver	1.5	n.s.	Ground	Seeds, insects
Cardinal Woodpecker	1.6	*	Trunks, branches	Arthropods
Crowned Hornbill	1.7	n.s.	Canopy	Fruit, insects
Bearded Woodpecker	2.1	*	Trunks, ground	Arthopods
Long-billed Crombec	2.3	*	Bushes	Insects
Fork-tailed Drongo	2.3	***	Air, ground	Insects
Red-billed Hornbill	3.4	***	Ground	Animals
Yellow-eyed Canary	4.7	n.s.	All levels	Seeds, insects
Red-billed Wood-hoopoe	10.5	***	Trunks, ground	Arthropods
White-headed Black Chat	10.7	***	Ground, trunks	Arthropods

* Mean ratio: 1, for all surveys. Negative values: commoner in the north; positive values: commoner in the south.
† Of differences in relative abundance in 1987 census, taking account of vegetation correlates with numbers; n.s. — not significant, * $p < 0.05$, ** < 0.01, *** < 0.001.

Box 5.3 Feeding behaviour and diet
R.J. Douthwaite and C.C.D. Tingle

Methods Gut contents of birds collected were preserved in 70% ethanol. Food remains were sorted and identified under a binocular microscope, with reference to the pitfall trap collections of invertebrates from mopane woodland. In general, identification was only possible to order or family, but certain insects could be identified to species.

Results During October to November the seven species were separated clearly by feeding site and method (Table 5.3), and less clearly by diet (Table 5.4). Two species, the White-headed Black Chat and White-browed Sparrow-weaver, often fed together, in open ground, but the sparrow-weaver was mainly granivorous[33], whereas the chat, like the five remaining species, was largely insectivorous.

Feeding sites of Red-billed Wood-hoopoe and White-headed Black Chat were sometimes treated, or accidently contaminated, with DDT. Of the remaining species, feeding sites of the White-browed Sparrow-weaver and Black Tit were generally closest to, and the Striped Kingfisher's most distant from, spray deposits.

Beetles predominated numerically in the guts of Black Tit, Chinspot Batis, White Helmet Shrike and Red-billed Wood-hoopoe. Weevils were particularly important to Black Tits, whereas Chinspots took more chrysomelids. Ants, especially *Camponotus* spp., [Formicidae: Formicinae] and *Pheidole* spp. [Myrmicinae] predominated in the diet of the White-Headed Black Chat in both July and November. However, termites, especially *Odontotermes* spp. [Isoptera: Termitidae], important in July, were largely replaced by beetles in November. Grasshoppers, mainly acridids, were the main component in the diet of the Striped Kingfisher.

Predators, potentially more heavily contaminated with residues, figured most prominently in the diets of White-headed Black Chat (ants) and Striped Kingfisher (mantids and arachnids).

Woodland Birds

Table 5.3 Feeding site and method

Species	Feeding sites and methods
Red-billed Wood-hoopoe	Crevices in tree trunks, large branches and fallen logs; gleaning.
White-headed Black Chat	Open ground and tree trunks; gleaning-pouncing.
Black Tit	Branches; gleaning.
Chinspot Batis	Tree canopy; hawking-gleaning.
White Helmet Shrike	All levels in woodland; gleaning-pouncing.
Striped Kingfisher	Ground in wooded grassland; pouncing.
White-browed Sparrow-weaver	Open ground in woodland; gleaning.

Table 5.4 Frequency (%) of invertebrates in bird guts

	Red-billed Wood-hoopoe		White-headed Black Chat				Black Tit		Chinspot Batis		White Helmet Shrike		Striped Kingfisher	
			July		November									
	Guts	Items	Guts	Items	Guts	Items	Guts	Items	Guts	Items	Guts	Items	Guts	Items
Total (n)	12	145	11	389	10	152	9	51	10	57	5	34	10	52
Isopoda					10	+								
Diplopoda					40	4							20	4
Chilopoda					10	+							30	6
Scorpionida	17	1	9	+										
Solifugae													40	9
Araneae	58	5	9	+	20	1	44	8	60	12			60	14
Blattodea	8	1									2	3		
Mantodea			9	+			22	4	10	2	40	6	40	12
Hemiptera	58	10	18	+	40	3			20	5	20	3	20	4
Homoptera			9	+					10	2			10	2
Heteroptera	58	10			40	3			10	2	20	6	10	2
Isoptera	42	29	91	40	30	5							10	4
Termitidae	42	29	54	33	10	+							10	4
Hodotermitidae			9	+	20	5								
Orthoptera	25	2			10	1			10	2	60	12	60	30
Lepidoptera	42	5	9	+			44	10	30	7				
larva	42	5	9	+					30	5				
Coleoptera	83	39	54	2	100	24	100	75	100	67	100	42	60	15
Caraboidea	8	1	9	+	40	4							10	2
Scarabeioidea					10	+	22	4					30	8
Histeridae					10	+								
Tenebrionidae					10	+								
Chrysomelidae	33	3					33	8	90	25				
Curculioniodea	25	2	18	+	60	8	56	31	10	2	20	18		
Diptera					10	+							10	2
Neuroptera (larva)	17	5			10	+								
Hymenoptera	17	3	91	57	100	56	22	4	30	5	80	30		
Formicidae	8	3	91	57	100	56					40	15		
Parasitica									10	2	20	6		
Sphecidae											20	3		
Scolioidea			9	+										

Note: all samples taken in October/November, except for one of White-headed Black Chat
+, < 1%

Box 5.4 DDT residues

Methods Samples of Red-billed Wood-hoopoe (RbWh), White-headed Black Chat (WhBC), Black Tit (BT), Chinspot Batis (CsB), Striped Kingfisher (SK), White Helmet Shrike (WHS) and White-browed Sparrow-weaver (WbSw) were shot in sprayed areas 1–3 and 12–17 months after treatment. Samples of White-headed Black Chat, Chinspot Batis and White-browed Sparrow-weaver were also taken from unsprayed woodland.

Carcasses, minus feathers, viscera, beaks, tarsi and feet, were stored in 10% formalin before being homogenized in a blender and then deep frozen, pending analysis for DDT residues. Carcass fat content and residues of o,p' and p,p' isomers of DDT, DDD and DDE were determined at the Natural Resources Institute[35], except for samples of White-headed Black Chat and Chinspot Batis collected in July 1987, which were analysed under contract by ADAS.[36] Analytical methods used by the two laboratories were similar. Residues were extracted with hexane in Soxhlet apparatus for 4–6 h. The extracts were concentrated and redissolved in cyclohexane/ethyl acetate. Extracts were cleaned at the Natural Resources Institute by gel permeation chromatography (GPC), elution with ether hexane in 10 g columns of florisil and elution with hexane in 10 g columns of 6% deactivated neutral alumina.[37] Clean-up by ADAS was by GPC and Sep-Pak cartridges. Clean extracts were examined by gas liquid chromatographs (GLC) fitted with electron capture detectors to determine the pesticide residues present. Confirmation of a small number of analyses was obtained by mass spectral detection. Both laboratories recovered over 85% of residues in sample checks.

Interpretation of results Unaltered DDT, DDD or DDE were found in every sample. The origin of DDD in wildlife specimens is controversial[38] but some, if not most, is derived from DDT post-mortem.[39,40] Conversion continues when tissue is stored in formaldehyde[41] but the amount converted equals about 1.11 times the level of DDD.[40] Corrected figures of DDT, shown as 'DDT', are therefore used below.

Technical DDT contains about 92% unaltered DDT, which degrades to DDE. The higher the proportion of unaltered DDT in ΣDDT in animal tissue, the closer the temporal, spatial or trophic distance to a spray deposit and the greater the risk, because unaltered DDT is more toxic than DDE. The effects of DDT and DDE may be additive[42], but in experimentally poisoned cowbirds, *Molothrus ater*, one part of DDT was equivalent to about 6.64 parts of DDE.[43,44] This ratio has been used to convert DDE residue levels to toxicological equivalents of DDT. The total number of toxicological equivalents of DDT is expressed as 'ΣDDT' in Figures 5.3–5.5. The distribution of residue levels in a large number of samples is almost invariably negatively skewed. Values were there-

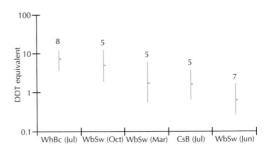

Figure 5.3 Residue levels in woodland species from unsprayed areas

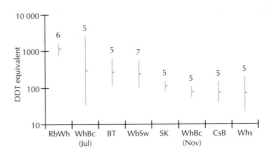

Figure 5.4 Residue levels in seven woodland species from areas sprayed 1–3 months previously

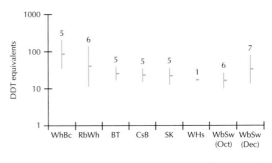

Figure 5.5 Residue levels in seven woodland species from areas sprayed 14–17 months previously

fore transformed logarithmically to facilitate statistical treatment; means are geometric and levels are given in ppm of extractable lipid. Sample variances were compared by the F test and means by the T test.

DDT residues dissolve in body fat and are harmless unless the fat is metabolized. Laboratory work has shown that residue levels in the brain or carcass lipid of birds killed by DDT are distinct from those surviving the experimental dose.[45] However, a range of environmental factors will determine whether critical levels are reached in the wild. The need to metabolize fat varies seasonally, with temperature and the availability of food, and differs between the sexes especially during the breeding season. Females 'dump' residues into the egg, just as lactating mammals shed burdens in milk. Body weight — and associated fatness — is a major factor in determining which birds of a given species die under pesticide exposure.[42] Heavy birds tend to lose more weight and die later than light birds. Drought, and its effects on the food supply, may therefore be a significant factor affecting risk in the present context. Previous exposure to DDT may also be important, as the ability to detoxify residues, which varies between species, is increased by previous exposure.

Conclusions The highest proportions of unaltered DDT (corrected for DDD) in ΣDDT were found in Red-billed Woodhoopoe (mean 77%), White-headed Black Chat (62%), Black Tit (52%) and White-browed Sparrow-weaver (50%) within 1–3 months of treatment, consistent with their feeding close to spray deposits. The lowest proportion at this time was found in Chinspot Batis (23%).

The lowest proportions of unaltered DDT were found in Striped Kingfisher (2%) and Chinspot Batis (7%) in October, 15–16 months after treatment. Levels in three sedentary species from the unsprayed area were indicative of windborne contamination. Residues in White-browed Sparrow-weavers contained 10–18% unaltered DDT in March, July and October, compared with 25% in Chinspot Batis and 52% in White-headed Black Chat in July.

The risk posed by residues varied by an order of magnitude between species within a treatment area at any one time (Figures 5.3–5.5). Residue burdens fell by an order of magnitude in the year following spraying operations (Figure 5.5). At any one time, exposure in birds in sprayed areas was about two orders of magnitude higher than in birds in unsprayed areas with background contamination. Exposure in the White-headed Black Chat 1–3 months after treatment varied by an order of magnitude dependent upon diet.

Effects of DDT on selected species

Seven common species, which could be readily sampled, were studied in detail to assess whether feeding habits and DDT residue accumulation could explain their relative abundance in the sprayed and unsprayed areas.

Red-billed Wood-hoopoe

The Red-billed Wood-hoopoe (Plate 18) is usually common where large trees occur, living in noisy territorial groups of a breeding pair and up to 15 non-reproductives, which help the pair raise young. Rainfall is the main determinant of breeding success and survival is closely related to territory quality. Tree cavities, where roosting and nesting birds are safe from predation by genets and driver ants, are the most important territorial resource. Both sexes, and especially the male, are strongly philopatric.[46]

Three surveys of occurrence were made in the study area, in July 1987, October 1988 and October 1989. After birds were located, pre-recorded calls were played back on a tape recorder to lure them closer for more accurate counting. In July 1987, flocks were significantly larger in the unsprayed area than in the area treated annually since 1984 (Mann & Whitney, $p = 0.003$), but sighting frequency was not recorded (Table 5.5). Flock size had altered little by October 1988 in either the unsprayed area or the area treated annually since 1984. Flock size in an area first treated in July 1987 remained similar to the unsprayed area, but flocks were encountered less often. One year later, in October 1989, flocks were fewer and smaller and the species was almost as scarce as in the area treated annually since 1984. By contrast, there had been little change in occurrence in the unsprayed area.

Birds, shot in October and November, two months after spraying ended, had fed exclusively upon arthropods, mainly beetles and termites (Table 5.4). These had been taken from crevices in tree trunks, large branches and from the ground (Table 5.3).

DDT residue levels in some of the birds were amongst the highest recorded for any species and there can be little doubt that the population decline after spraying was due to a lethal accumulation of DDT residues (Figure 5.4).

Table 5.5 Decline of Red-billed Wood-hoopoes in treated area at Siabuwa

	Survey		
	1987	1988	1989
Treated 1984–1989			
Median flock size	4	3	3
Mean time between sightings (min)	–	186	174
Treated 1987–1989			
Median flock size	7–8	7	4
Mean time between sightings (min)	–	54	114
Untreated			
Median flock size	7–8	7	7–8
Mean time between sightings (min)	–	36	48

White-headed Black Chat

The White-headed Black Chat (Plate 19) is an arboreal thrush, virtually endemic to mopane and miombo woodland in southern Africa.[48] It lives in groups of 1–6, breeds co-operatively, and feeds largely on arthropods taken from the ground or from tree trunks. The species is resident but the degree of philopatry is unknown. In the Zambezi valley, it is common and conspicuous in old woodland with tall, bare-limbed trees, an open canopy, little understorey and patchy ground cover. Territory size at Siabuwa was estimated to be 5–25 ha.

The species is territorial and birds responded conspicuously to playback of their song on a tape recorder. Surveys of relative abundance were made in treated and untreated woodland in six widely separated areas of north-west Zimbabwe. Numbers were invariably lower in the treated area, than in the adjacent untreated area even when no treatment had been applied for seven years.[34]

Population changes at Siabuwa were monitored from July 1987 to January 1991 and related to spray treatments (Figure 5.7). In an area treated in 1987, 1988 and 1989 numbers fell by 88% over 33 months following the first spray, mainly due to a reduction in the number of occupied sites (Figure 5.6). Numbers in the unsprayed area fell by 13% during the same period. In January 1991, numbers were increasing again in areas treated in 1987 and in 1987–89. Isolated groups were also present in an area treated six times between 1984 and 1989.

In a replicate study, numbers also declined in the Omay Communal Area following treatment with DDT.[49]

Food supply, breeding success and DDT residue accumulation were investigated as possible causes of decline at Siabuwa.[47]

Ants, especially *Camponotus* spp., were numerically predominant in the diet in both July and November. Termites, which were also important in July, were largely replaced by beetles in November (Table 5.4).

Ant and termite activity at sprayed sites in the study area was as great as, or greater than that at unsprayed sites. Ants (*Camponotus* spp.) from sprayed sites held mean

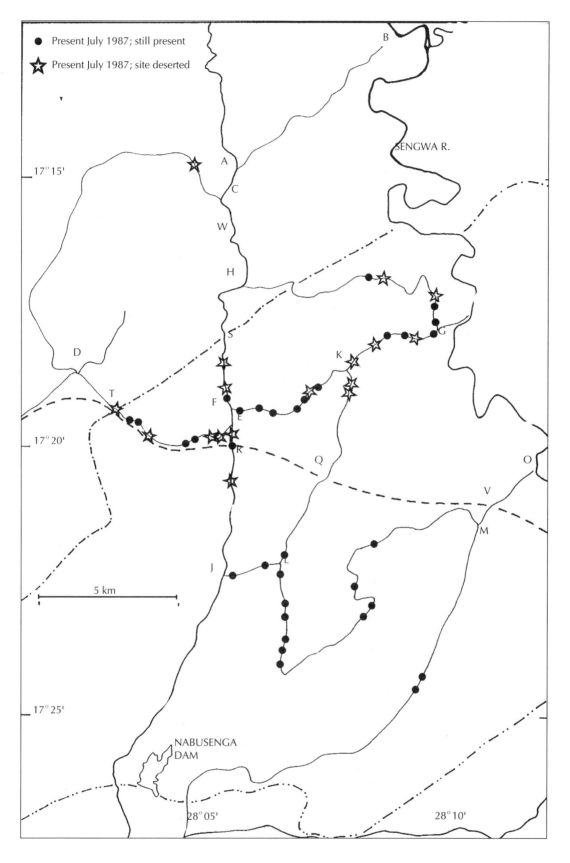

Figure 5.6 Distribution of White-headed Black Chats at Siabuwa in July 1987 and site desertions by October 1988

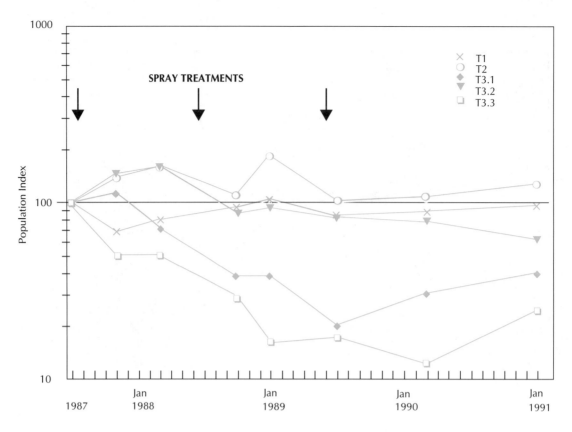

Figure 5.7 Change in White-headed Black Chat numbers at Siabuwa. Numbers in consecutive surveys are related by a chain index[6]. Unsprayed (T1); Sprayed: edge of 1987 and 1988 treatments (T2); 1987 (T3.1); 1987 and on edge of 1988 and 1989 (T3.2); 1987, 1988 and 1989 (T3.3).

residue levels of 8.7 ppm (max. 218 ppm) dry wt. ΣDDT, of which 67% was unaltered DDT. Termites and beetles had mean residue levels of 3.3 ppm (max. 14 ppm) and 0.9 ppm (max. 8 ppm) ΣDDT, of which 44% and 37% was unaltered DDT respectively.

Fledging success of White-headed Black Chats in adjacent sprayed and unsprayed areas was similar. In November 1987, fledged young were seen in 10 out of 14 groups in the area sprayed for the first time earlier that year, and in three groups found in the area treated annually since 1984. After breeding in 1988, young were seen in 38% of groups ($n = 21$) in the area first treated in 1987, and in 36% of groups ($n = 28$) in the unsprayed area.

Residues of DDT, DDD and DDE were found in all 23 carcasses examined. Birds collected in the dry season (July) from an area sprayed one month before contained up to 2206 ppm DDT, 367 ppm DDD and 578 ppm DDE extractable lipid (86, 17 and 27 ppm dry weight respectively) and there can be little doubt that these would have proved fatal for some. Residue levels in birds collected in the early rains (November) from a recently sprayed area in Omay were much lower, probably because birds were feeding largely on beetles, especially weevils, instead of ants and termites. Although there was little risk of poisoning at the time, the population subsequently declined, probably due to re-exposure to high residue levels with another seasonal shift in diet.

It was concluded that spraying operations with DDT have had a severe, and perhaps prolonged, impact on the White-headed Black Chat population of north-west Zimbabwe. The initial decline at Siabuwa was due to a lethal accumulation of DDT residues from prey, especially *Camponotus* spp. ants, rather than from reduced availability of prey or lower fledging success, but why numbers should continue to decline for 2–3 years after spraying ended is unclear. However, evidence suggests some ant species remain heavily contaminated with DDT two years after treatment.

Black Tit

The Black Tit is a common, widespread resident of the moister woodlands of southern Africa. It is philopatric, living in groups of 2–5 birds which breed co-operatively in territories of 25–48 ha.[50]

Numbers in the north and south of the study area were similar (Table 5.2). In October, birds had fed largely on beetles gleaned from tree branches, mainly found above the target zone for DDT applications (Tables 5.3–5.4). DDT residue levels generally were significantly below those found in Red-billed Wood-hoopoes and no effect of exposure was apparent (Figures 5.4–5.5). The most heavily contaminated bird, shot one month after spraying ended, contained 935 ppm DDT, 176 ppm DDD and 432 ppm DDE ppm extractable lipid (= 1195 ppm DDT equivalents), half the amount found in the most heavily contaminated Red-billed Wood-hoopoe and White-headed Black Chat.

White-browed Sparrow-weaver

The White-browed Sparrow-weaver is a sedentary species, living in groups comprising a breeding pair and up to nine non-reproductives. These breed co-operatively and defend territories some 50 m in diameter centred on a conspicuous group of nests built of grass in a tree (Figure 5.8).[1,51–54] The White-browed Sparrow-weaver was the most abundant species in the study area accounting for about 25% of all birds.

Point count surveys indicated more birds were present in the south of the study area than in the north. This was confirmed by a survey of colonies made in December 1989. The presence or absence of colonies in plots 100 m in diameter was recorded at regular intervals along tracks through the study area. Colony density was estimated as 60/km^2 in the south and 48/km^2 in the north. However, colonies were restricted to

Figure 5.8 Nests of White-browed Sparrow-weavers were conspicuous and their abundance in sprayed and unsprayed areas easily assessed (*Source*: H.Q.P. Crick)

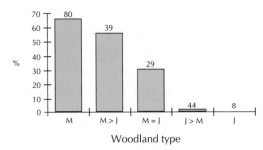

Figure 5.9 Percentage of woodland plots with White-browed Sparrow-weaver colonies (number of sample plots indicated at head of histogram)

mopane woodland which was more widespread in the south (Figure 5.9). Estimated densities in mopane woodland in the north and south were 74 and 71 colonies/km², respectively.

Colony size varied with area and time, but not with spray treatment. In 1987 colonies in the north were twice the size of those in the south (means 15.5 and 8 nests/colony), probably due to more severe drought in the south. Two years later, following good rains, average colony size in the north and south was similar (16 and 17 nests/colony respectively). It was concluded that the White-browed Sparrow-weaver population was unaffected by spray treatments.

Residue levels in birds sampled at intervals from 3 to 17 months after spraying were highest at 3 months, falling almost to background levels 12 months after treatment (Figures 5.3–5.5). The most heavily contaminated bird contained 1083 mg DDT-equivalents/kg carcass lipid, less than half the amount found in White-headed Black Chats and Red-billed Wood-hoopoes, whose populations collapsed after spraying. White-browed Sparrow-weavers and White-headed Black Chats occur together and both species feed on the ground. Their differing exposure to DDT is probably related to diet. The White-browed Sparrow-weaver eats some arthropods, especially termites, but over 90% of the items eaten were seeds.[33] The White-headed Black Chat, on the other hand, feeds entirely on invertebrates, especially ants.

Striped Kingfisher

The Striped Kingfisher breeds co-operatively and males predominate in the population.[55] It occurs in open wooded grassland, pouncing from a perch onto prey, which in October consisted largely of grasshoppers, spiders and solifugids (Tables 5.3–5.4).

During most of the study Striped Kingfishers appeared equally abundant in the north and south of the study area. However, in October 1989, kingfishers were much more conspicuous, and were especially noisy, in the north (Figure 5.10). Breeding was probably in progress and all three females shot in the the sprayed area had large oocytes (< 3–6 mm in diameter) and enlarged oviducts. None of the seven males, on the other hand, had enlarged testes (< 2–3 mm in length). The difference in apparent numbers between the north and south of the study area at this time cannot be explained satisfactorily but may have been due to local suitability for breeding. An effect of spraying seems unlikely given the diet and relatively low residue burden found in birds from sprayed areas (Figure 5.4).

Chinspot Batis

The Chinspot is a common flycatcher in the open, deciduous woodlands of southern Africa where it captures insects from the tree canopy by hawking-gleaning.

Woodland Birds

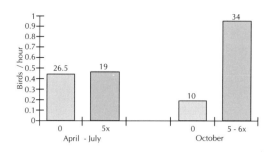

Figure 5.10 Sighting rates of Striped Kingfisher in sprayed and unsprayed areas during April–July and October 1989 (hours of observation indicated at head of histogram; 0, 5x, 5–6x, number of spray treatments)

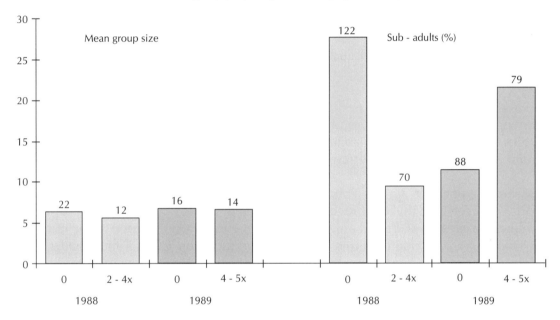

Figure 5.11 Mean flock size and proportion (%) of sub-adult White Helmet Shrikes in sprayed and unsprayed area populations, in April–May 1988 and May 1989 (sample sizes indicated at head of histogram; 0, 2–4x, 4–5x, number of spray treatments)

Numbers in the north and south of the study area were similar (Table 5.2). In October, birds had fed largely on small beetles (Table 5.4) and their DDT residue burden was relatively low; there was little risk numbers would be adversely affected by spraying operations (Figure 5.4).

White Helmet-shrike

The White Helmet Shrike is resident and territorial, occurring in flocks of 2–17 birds and breeding co-operatively.

The relative abundance of White Helmet Shrikes in the north and south of the study area varied widely between surveys (Figure 5.11). A more intensive study made in May 1989, using the playback of tape-recorded calls, showed that there was little difference between areas: flocks were sighted on average every 24 min., or 2.6 playback attempts in the north, compared with 28 min., or 3.3 playback attempts in the south ($n = 14$ flocks in each area).

Flock size and age structure were recorded in April–May 1988 and May 1989 (Figure 5.11). Variation in flock size and proportion of young birds was greater between years than between spray treatments.

White Helmet Shrikes feed at all levels, but especially low down, gleaning foliage for beetles, ants and grasshoppers (Tables 5.3 and 5.4). DDT residue burdens were lower than in any other species and no effect of spraying was evident (Figures 5.4 and 5.5).

Conclusions

Exposure to DDT residue levels in the species studied was related mainly to feeding station and diet. Highest levels were found in Red-billed Wood-hoopoes and White-headed Black Chats, both of which eat arthropods from tree trunks likely to be sprayed with DDT. Both also roost and nest in tree holes at risk of treatment. Intermediate residue levels were found in the Black Tit, which takes arthropods mostly from branches above the spray target, and in the White-browed Sparrow-weaver, which often feeds alongside the White-headed Black Chat, but eats seeds rather than arthropods. Lower levels were found in species feeding on arthropods associated with grassland (Striped Kingfisher) and foliage of trees and bushes (Chinspot Batis and White Helmet Shrike).

The Red-billed Wood-hoopoe and White-headed Black Chat declined in the wake of spraying operations but numbers of the other five species studied in detail did not vary with treatment. Rainfall and habitat differences were probably more important than residue contamination in determining abundance.

The number of other species eating heavily contaminated arthropods is probably small. However, both the Red-billed Hornbill and Fork-tailed Drongo are opportunists and likely to eat spray-affected animals: both were much less common in the north of the study area and numbers may have been reduced by DDT. Four woodpecker species, Bearded, Cardinal, Bennett's and Golden-tailed, also feed on arthropods from tree trunks. Their numbers in sprayed and unsprayed areas appeared similar, although the 1987 census results, corrected for differences in vegetation, indicated Bearded and Cardinal woodpeckers were significantly less common in the sprayed area. Two resident doves, the Cape Turtle Dove and Emerald-spotted Wood-dove, and one immigrant species, the Laughing Dove, first recorded after the good rains of 1987/88, were much less common in the north of the study area and may have been affected by DDT, although seeds comprise a large part of their diet. The Arrow-marked Babbler and Red-capped Robin, which feed on invertebrates in riverine thickets, were uncommon and only seen in unsprayed areas.

In conclusion, populations of two common woodland songbirds were severely affected, but not exterminated, by spraying operations. Several other species were probably affected by ground-spraying but it is clear that rainfall and habitat are more important determinants of bird numbers generally. The population dynamics of affected species and their long-term exposure to DDT residues have not been studied, making prognoses for population recovery uncertain. The persistence of DDT in ants, and relative scarcity of White-headed Black Chats in superficially suitable woodland seven years after treatment, suggest recovery in this species may take ten years or more.

Effects on Birds of Prey

African Goshawk

The study, begun in 1988, was a co-operative venture with the Zimbabwe Falconers' Club, Falcon College and Peterhouse School.

The African Goshawk (Plate 20) is a common raptor in riverine woodland, preying mainly on birds. Pairs at Siabuwa occupied small home ranges and bred between November and January.

Searches for nests were made through thick bush along streams during the breeding seasons of 1988/89, 1989/90 and 1990/91. By January 1991, only two of the ten known sites in the north of the study area remained in use, one in an area treated annually since 1984 and the other in an area treated since 1987 (Table 5.6). At least four sites were deserted during the study period. By contrast, nine of the 11 sites known in the southern, unsprayed area were still in use, and not more than two were abandoned during the study period.

Table 5.6 Status of African Goshawk nest sites in January 1991

Area	Nest sites	Used in 1990/91	Deserted during the study
Treated 3–6 x	10	2	2
Treated 1 x (1987)	5	3	2
Untreated	11	9	1

Eggs were collected from seven nests and shell fragments from another three. Egg contents were analysed for DDT residues and the thickness of eggshells measured using a microscope stage micrometer accurate to ±0.004 mm. Clutch means are used in calculations where more than one egg was collected from a nest.

Both ΣDDT and DDE content were related to number of past spray treatments (Table 5.7). High residue levels were found in heavily treated areas and low levels in unsprayed areas. Eggshell thickness was correlated significantly with DDE content ($r_s = -0.72$, $p < 0.05$) (Figure 5.12). Eggshells from heavily sprayed areas averaged 22% thinner than shells from unsprayed areas and addled eggs, with cracked shells, were found in nests in both heavily sprayed and less sprayed areas (Figure 5.13).

Remains of 27 prey items were recovered and identified. Nineteen (70%) were birds, five (18%) mammals and three (11%) lizards. Significantly, remains of Red-billed Wood-hoopoe and other species suspected of accumulating high residue levels (e.g.

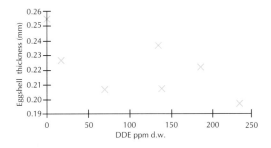

Figure 5.12 Eggshell thickness versus DDE content of African Goshawk eggs

DDT in the Tropics

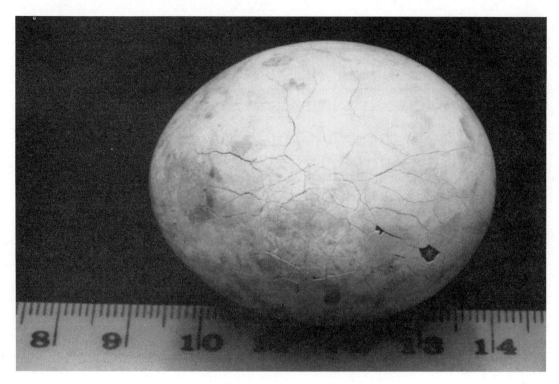

Figure 5.13 Cracks in an African Goshawk egg were indicative of extreme shell thining due to high DDE residues in the female (*Source*: R.J. Douthwaite)

Table 5.7 DDT residue levels (geometric mean and range ΣDDT ppm dry weight) and eggshell thickness of African Goshawk eggs collected at Siabuwa, 1988–91

Previous treatments	ΣDDT			DDE Mean %	Eggshell thickness (mm)		
	(n)	Mean	(range)		(n)	Mean	% thinning
0	1	1	–	100	2	0.2545	–
1–2	4	88	18–174	80	5	0.220	14
5–6	2	273	229–326	75	3	0.1985	22

Red-billed Hornbill and doves) were found, demonstrating a short food chain between the spray deposit via bark, invertebrates and insectivorous bird, to avian predator. There can be little doubt that breeding failure and population decline at Siabuwa occurred as a result of DDE contamination arising from tsetse spraying operations.

The population dynamics of the African Goshawk have not been studied sufficiently to make a confident prognosis for recovery, but given the relatively rapid disappearance of DDT residues from the physical environment and potential prey species, recovery may be faster than that of the goshawk's counterpart, the European Sparrowhawk, which has taken 20 years or more to recover in the areas worst affected by diedrin and DDT.

Peregrine Falcon

A separate national survey of breeding success and DDT contamination in the Peregrine Falcon[32] found little evidence of failure (Box 5.5). However, residue levels sufficient to cause significant eggshell thinning were found in an egg from the Chizarira National Park, sprayed for the third and last time in 1984, suggesting tsetse spraying operations are likely to cause local breeding failure in this species.

Box 5.5 DDT residues and status of the Peregrine falcon in Zimbabwe
R. Hartley

In 1990, the Zimbabwe Falconer's Club, Department of National Parks and Wildlife Management and US Peregrine Fund Inc., assessed the impact of DDT on breeding success in the Peregrine Falcon.[32]

Single intact eggs were removed under licence from 11 sites, and eggshells from recent hatchings from another two. Seven of the sites were in areas treated with DDT for tsetse fly control. Residues analysis was carried out at the Institute of Terrestrial Ecology, Monks Wood. Every egg contained DDE, but only one, from Chizarira National Park contained enough (22 ppm wet weight) to cause significant eggshell thinning. The adjacent area was last treated in 1984, for the third time. The next highest level (8 ppm) was found in an egg from the Chirisa Safari Area nearby. Measurement of eggshell thickness remains to be done.

The number of known Peregrine eyries in Zimbabwe has increased from 10 in 1971, to 47 in 1984, 107 in 1990 and 116 in 1991, doubtless reflecting successful development of the ZFC database rather than a change in Peregrine population. Similarly, the number of known Taita Falcon eyries, Zimbabwe's rarest breeding falcon, has increased from 2 in 1983 to 16 in 1991.

Scientific names of birds mentioned in the text

African Golden Oriole *Oriolus auratus*
African Goshawk *Accipiter tachiro*
American Robin *Turdus migratorius*
Arrow-marked Babbler *Turdoides jardineii*
Bearded Woodpecker *Thripias namaquus*
Bennett's Woodpecker *Campethera bennettii*
Black Sparrowhawk *Accipiter melanoleucus*
Black Tit *Parus niger*
Black-crowned Tchagra *Tchagra senegala*
Black-eyed Bulbul *Pycnonotus barbatus*
Blue Waxbill *Uraeginthus angolensis*
Cape Turtle Dove *Streptopelia capicola*
Cardinal Woodpecker *Dendropicos fuscescens*
Chinspot Batis *Batis molitor*
Crested Barbet *Trachyphonus vaillantii*
Crowned Hornbill *Tockus alboterminatus*
Darter *Anhinga rufa*
Emerald-spotted Wood Dove *Turtur chalcospilos*
European Sparrowhawk *Accipiter nisus*
Fish Eagle *Haliaeetus vocifer*
Fork-tailed Drongo *Dicrurus adsimilis*
Golden-breasted Bunting *Emberiza flaviventris*
Golden-tailed Woodpecker *Campethera abingoni*
Grey Heron *Ardea cinerea*
Grey Hornbill *Tockus nasutus*
Grey-headed Sparrow *Passer griseus*
Lanner Falcon *Falco biarmicus*

Laughing Dove *Streptopelia senegalensis*
Long-billed Crombec *Sylvietta rufescens*
Pel's Fishing Owl *Scotopelia peli*
Peregrine Falcon *Falco peregrinus*
Puffback Shrike *Dryoscopus cubla*
Racket-tailed Roller *Coracias spatulata*
Red-billed Hornbill *Tockus erythrorhynchus*
Red-billed Wood-hoopoe *Phoeniculus purpureus*
Red-capped Robin *Cossypha natalensis*
Rock Bunting *Emberiza tahapisi*
Scimitar-billed Wood-hoopoe *Rhinopomastus cyanomelas*
Short-tailed Cisticola *Cisticola fulvicapilla*
Striped Kingfisher *Halcyon chelicuti*
Tawny-flanked Prinia *Prinia subflava*
Three-streaked Tchagra *Tchagra australis*
Wahlberg's Eagle *Aquila wahlbergi*
White Helmet Shrike *Prionops plumata*
White-browed Sparrow-weaver *Plocepasser mahali*
White-headed Black Chat *Thamnolaea arnoti*
Yellow-eyed Canary *Serinus mozambicus*
Yellow-throated Sparrow *Petronia superciliaris*

REFERENCES

1. Risebrough, R.W. (1986) Pesticides and bird populations. *Current Ornithology* **3**: 397–427.

2. Peakall, D.B. (1985) Behavioural responses of birds to pesticides and other contaminants. *Residue Reviews*, **96**: 45–77.

3. Hotchkiss, N. and Pough, R.H. (1946) Effect on forest birds of DDT used for Gypsy Moth Control in Pennsylvania. *Journal of Wildlife Management*, **10**: 202–207.

4. Marchant, J.H., Hudson, R., Carter, S.P. and Whittington, P. (1990) *Population trends in British breeding birds*. Tring: British Trust for Ornithology.

5. Cramp, S. and Conder, P.J. (1961) *The Deaths of Birds and Mammals connected with Toxic Chemicals in the first half of 1960*. Report No. 1 of the BTO-RSPB Committee on Toxic Chemicals. London: RSPB.

6. Cramp, S., Conder, P.J. and Ash, J.S. (1962) *Deaths of birds and mammals from toxic chemicals, January-June 1961*. Report No. 2 of the BTO-RSPB Committee on Toxic Chemicals. Sandy: RSPB.

7. Wallace, G.J., Nickell, W.P. and Bernard, R.F. (1961) *Bird Mortality in the Dutch Elm Disease Program in Michigan*. Bulletin 41. Bloomfield Hills, Michigan: Cranbrook Institute of Science.

8. Hickey, J.J. and Hunt, L.B. (1960) Initial songbird mortality following a Dutch Elm disease control program. *Journal of Wildlife Management*, **24**: 259–265.

9. Hickey, J.J. and Hunt, L.B. (1960) Songbird Mortality following annual programs to control Dutch Elm disease. *Atlantic Naturalist*, **15**: 87–92.

10. Hunt, L.B. (1960) Songbird breeding populations in DDT-sprayed Dutch Elm disease communities. *Journal of Wildlife Management*, **24**: 139–146.

11. Mehner, J.F. and Wallace, G.J. (1959) Robin populations and insecticides. *Atlantic Naturalist*, **14**: 4–10.

12. Robbins, C.S., Springer, P.F. and Webster, C.G. (1951) Effects of five-year DDT application on breeding bird population. *Journal of Wildlife Management*, **15**: 213–216.

13. Wurster, C.F., Wurster, D.H. and Strickland, W.N. (1965) Bird Mortality after Spraying for Dutch Elm Disease with DDT. *Science*, **148**: 90–91.

14. Henny, C.J., Blus, L.J., Krynitsky, A.J. and Bunck, C.M. (1984) Current impact of DDE on Black-crowned Night-herons in the intermountain West. *Journal of Wildlife Management*, **48**: 1–12.

15. Ratcliffe, D.A. (1967) Decrease in eggshell weight in certain birds of prey. *Nature*, **215**: 208–210.

16. Wiemeyer, S.N., Lamont, T.G., Bunck, C.M., Sindelar, C.R., Gramlich, F.J., Fraser, J.D. and Byrd, M.A. (1984) Organochlorine pesticide, polychlorbiphenyl, and mercury residues in Bald Eagle eggs — 1969–1979 — and their relationships to shell thinning and reproduction. *Bulletin of Environmental Contamination and Toxicology*, **13**: 529–549.

17. Nisbet, I.C.T. (1988) Organochlorines, reproductive impairment, and declines in Bald Eagle populations: mechanisms and dose-response relationships. In: *Raptors in the Modern World* (Meyburg, B.-U. and Chancellor, R.D., eds.) pp. 483–490. Berlin: Weltarbeitsgruppe für Greifvogel und Eulen e. V.

18. Peakall, D.B., Cade, T.J. White, C.M. and Haugh, J.R. (1975) Organochlorine residues in Alaskan Peregrines. *Pesticides Monitoring Journal*, **8**: 255–260.

19. Blus, L.J., Belisle, A.A. and Prouty, R.M. (1974) Relationships of the Brown Pelican to certain environmental pollutants. *Pesticide Monitoring Journal*, **7**: 181–194.

20. Blus, L.J. (1982) Further interpretation of the relation of organochlorine residues in Brown Pelican eggs to reproductive success. *Environmental Pollution, Ser. A*, **28**: 15–33.

21. Fyfe, R.B., Risebrough, R.W., and Walker, W. (1976) Pollutant effects on the reproduction of the Prairie falcons and Merlins of the Canadian Prairies. *Canadian Field-Naturalist*, **90**: 346–355.

22. Spitzer, P.R., Risebrough, R.W., Walker, W. II, Hernandez, R., Poole, A., Puleston, D. and Nisbet, I.C.T. (1978) Productivity of Ospreys in Connecticut-Long Island increase as DDE residues decline. *Science*, **202**: 333–335.

23. Newton, I. (1979) *Population Ecology of Raptors*. Berkamsted: Poyser.

24. Irwin, M.P.S. (1981) *The Birds of Zimbabwe*. Harare: Quest.

25. Hustler, K. (1992) in litt.

26. Attwell, S. (1981) in litt.

27. Conway, A.J. (1981) in litt.

28. Maasdorp, R.A.C. and Zyl, M. van. (1987) Possible side-effects of insecticides. *Honeyguide*, **33**: 154.

29. Thomson, W.R. (1981) Letter. *Herald newspaper*, 30 April 1981, Harare.

30. Matthiessen, P. (1985) Contamination of wildlife with DDT insecticide residues in relation to tsetse fly control operations in Zimbabwe. *Environmental Pollution Ser. B*, **10**: 189–211.

31. Taylor, R.D. and Fynn, K.J. (1978) Fish eagles on Lake Kariba *Rhodesia Science News*, **12**: 2.

32. Hartley, R. (1991) The Zimbabwe Falconers Club — current research and conservation projects. *Endangered Wildlife*, **8**: 9–15.

33. Douthwaite, R.J. (1992) Effects of DDT treatments applied for tsetse fly control on White-browed Sparrow-weaver (*Plocepasser mahali*) populations in NW Zimbabwe. *African Journal of Ecology*, **30**: 233–244.

34. Douthwaite, R.J. (1992) Effects of DDT treatments applied for tsetse fly control on White-headed Black Chat (*Thamnolaea arnoti*) populations in Zimbabwe. Part I: Population changes. *Ecotoxicology*, **1**: 17–30.

35. Natural Resources Institute, (1990) *DDT and Deltamethrin in Avian Tissue*. Pesticide Residues Analysis Report No. 89BY. Chatham; UK: Natural Resources Institute.

36. ADAS (1990) *Determination of Organochlorine Pesticide residues in Avian Tissue Samples*. Contract Report No. C/89/0059. Tolworth UK: Agricultural Development and Advisory Service.

37. Telling, G.M., Sissons, D.J. and Brinkman, H.W. (1977) Determination of organochlorine insecticide residues in fatty foodstuffs using a clean-up technique based on single column of activated alumina. *Journal of Chromatography*, **137**: 405–423.

38. Bailey, S., Bunyan, P.J., Rennison, B.D. and Taylor, A. (1969) The metabolism of 1,1-Di(p-chlorophenyl)-2,2,2-trichloroethane and 1,1-Di(p-chlorophenyl)-2,2-dichloroethane in the pigeon. *Toxicology and Applied Pharmacology*, **14**: 13–22.

39. Barker, P.S. and Morrison, F.O. (1964) Breakdown of DDT to DDD in mouse tissue. *Canadian Journal of Zoology*, **42**: 324–325.

40. Jefferies, D.J. and Walker, C.H. (1966) Uptake of pp'-DDT and its Post-mortem Breakdown in the Avian Liver. *Nature*, **212**: 533–534.

41. French, M.C. and Jefferies, D.J. (1971) The Preservation of Biological Tissue for Organochlorine Insecticide Analysis. *Bulletin of Environmental Contamination and Toxicology*, **6**: 460–463.

42. Stickel, W.H., Stickel, L.F. and Coon, F.B. (1970) DDE and DDD Residues Correlated with Mortality of Experimental Birds. *Pesticide Symposia*. Miami: Halos Associates.

43. Stickel, L.F., Stickel, W.H. and Christensen, R. (1966) Residues of DDT in Brains and Bodies of Birds That Died on Dosage and in Survivors. *Science*, **151**: 1549–1551.

44. Stickel, W.H., Stickel, L.F., Dyrland, R.A. and Hughes, D.L. (1984) DDE in Birds: Lethal Residues and Loss Rates. *Arch. Environmental Contamination and Toxicology*, **13**: 1–6.

45. Stickel, L.F. (1973) Pesticide Residues in Birds and Mammals. In: *Environmental Pollution by Pesticides*. (Edwards, C.A., ed.) London: Plenum Press.

46. Ligon, J.D. and Ligon, S.H. (1989) Green Woodhoopoe. In: *Lifetime Reproduction in Birds*. (Newton, I., ed.). London: Academic Press.

47. Douthwaite, R.J. and Tingle, C.C.D. (1992) Effects of DDT treatments applied for tsetse fly control on White-headed Black Chat (*Thamnolaea arnoti*) populations in Zimbabwe. Part II: Cause of decline. *Ecotoxicology*, **1**: 101–115.

48. Benson, C.W., Brooke, R.K., Dowsett, R.J. and Irwin, M.P.S. (1971) *The Birds of Zambia*. Glasgow: Collins.

49. Lambert, M.R.K., Grant, I.F., Smith, C.L., Tingle, C.C.D. and Douthwaite, R.J. (1991) *Report on the Effects of Residual Deltamethrin Ground-Spraying on Non-target Wildlife*. Chatham, UK: Natural Resources Institute.

50. Tarboton, W.R. (1981) Co-operative breeding and group territoriality in the Black Tit. *Ostrich*, **52**: 216–225.

51. Collias, N.E. and Collias, E.C. (1978a) Co-operative breeding behaviour in the White-browed Sparrow Weaver. *Auk*, **95**: 472–484.

52. Collias, N.E. and Collias, E.C. (1978b) Survival and intercolony movement of White-browed Sparrow Weavers *Plocepasser mahali* over a two-year period. *Scopus*, **2**: 75–76.

53. Lewis, D. (1982a) Co-operative breeding in a population of White-browed Sparrow Weavers. *Ibis*, **124**: 511–522.

54. Lewis, D. (1982b) Dispersal in a population of White-browed Sparrow Weavers. *Condor*, **84**: 306–312.

55. Fry, C.H., Keith, S. and Urban, E.K. (1988) *The Birds of Africa*. Vol. 3. London: Academic Press.

56. Newton, I. (1986) *The Sparrowhawk*. Calton: Poyser.

57. Newton, I. and Wyllie, I. (1992). Recovery of a sparrowhawk population in relation to declining pesticide contamination. *Journal of Applied Ecology*, **29**: 476–484.

58. Douthwaite, R.J. (In press) Occurrence and consequences of DDT residues in woodland birds following tsetse fly spraying operations in NW Zimbabwe. *Journal of Applied Ecology*.

59. Hartley, R.R. and Douthwaite, R.J. (1994) DDT residues and the effect of DDT spraying to control tsetse flies on an African Goshawk population in the Zambezi Valley, Zimbabwe. *African Journal of Ecology*, **32**: 000–000.

60. James, F.C. and Shugart, H.H. (1970) A quantitative method of habitat description. *Audubon Field Notes*, **24**: 727–736.

6

NOCTURNAL ANIMALS

A N McWilliam

Bats

Bats have long been regarded as particularly susceptible to poisoning from organochlorine insecticides and a considerable body of evidence has implicated the agricultural use of DDT in the massive decline of certain cave populations, particularly in the cotton-growing areas of the United States.[1,2] In Britain, bats were frequently more contaminated with DDT residues than either insectivorous or carnivorous birds in an area where the insecticide was frequently applied in a farming context.[3] In the only detailed study to date that has monitored bats as non-target animals to DDT-spraying operations, five species were sampled in Dougas fir forests in north America subject to outbreaks of tussock moth.[4] Although immediate post-spray residues were not assessed, none of the individuals were at risk a year after spraying, when total DDT levels averaged 6.9 ppm wet weight in the most contaminated species.

Several factors promote bio-accumulation of pesticides in bats, particularly those such as DDT and its metabolites that are lipophilic, as many species experience pronounced fat cycles related to reproduction and migration. With their relatively high metabolic rates and a food intake of up to their own body weight of insects a night, bats are efficient integrators of insecticides. Exposure to harmful chemicals is further increased by their longevity, and any mortality magnified at a population level by low reproductive and recruitment rates. In addition, bats are also sensitive to indirect effects of spraying, such as any DDT-induced reduction in insect abundance, potentially leading to a reduction in reproductive success through poor body condition. Finally, many species are colonial and live in large tree or cave roosts where they are vulnerable to direct application, such as occurs during spraying in tsetse control operations. Thus, substantial mortality of both juveniles and adults was still being recorded at a nursery roost of little brown bats in the USA some two years after treatment with DDT.[5]

Under conditions of chronic exposure, sub-lethal effects of DDT have been documented for bats and other mammals which may contribute to population declines through reduction in individual reproductive success. For example, a halving of ATPase activity occurred in bat brains subject to concentrations of DDT and DDE much lower even than normally found in free-living bats.[6] A depression of an immune response was found in mice exposed to sub-toxic doses of DDT that increased in both a dose- and time-dependent fashion[7] and young of female mice dosed with only 2.5 mg/kg of DDT were slow to acquire a conditioned avoidance response.[8]

Significantly, a single low dose of DDT to neonatal mice, amounting to only 0.5 ppm by body weight, leads to a permanent hyperactive condition in the mice as adults.[9] Animal behaviour can be disturbed by DDT-induced hyper- or hypothyroidism[10] and DDT has also been shown to induce the mixed function oxidase enzymes involved in the metabolism of steroid hormones.[11]

Finally, DDT-dosed bats lost weight more rapidly than controls[12,13] and an increase in the metabolic rate of shrews fed on a DDE-dosed diet was demonstrated by Braham and Neal.[14]

For investigations among mammals at the community level, bat faunas are also good indicator groups because of their relative species richness and diversity of foraging guilds. Consequently, as the mammalian order second only to rodents in the number of species, bats are becoming a major focus of conservation effort. In Zimbabwe, there are about 60 species[15,16] and the studies of Fenton and co-workers at the Sengwa Wildlife Research Area have established that many of these species are amenable to ecological investigation.[17–19]

In an initial survey of wildlife contamination by DDT spraying against tsetse fly in Zimbabwe, Mathiessen[20] found that residues accumulated in *Scotophilus borbonicus*, a common bat chosen as an indicator species, although not to such a level as to cause acute poisoning. Thus, because of the high diversity and relative abundance of a potentially susceptible bat fauna (Table 6.1; Figure 6.1, 6.2) it was decided to monitor the response of this group to exposure from DDT-spraying operations at individual, population and community levels.

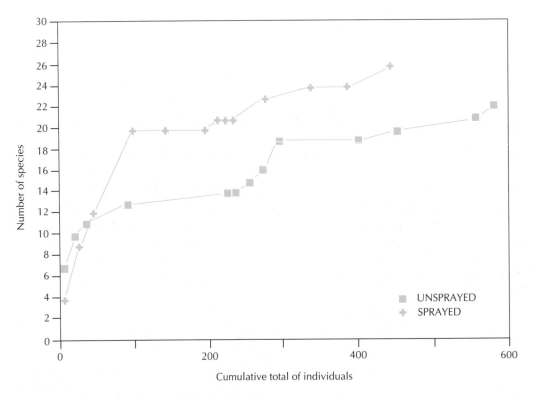

Figure 6.1 Bat species diversity in sprayed and unsprayed areas

Table 6.1 Bat species captured at Siabuwa, ranked by overall abundance.

Rank	Species		Unsprayed	Sprayed
1	*Nycticeius schlieffenii*	(V)	+	+
2	*Eptesicus somalicus*	(V)	+	+
3	*Scotophilus borbonicus*	(V)	+	+
4	*Hipposideros caffer*	(H)	+	+
5	*Pipistrellus rueppellii*	(V)	+	+
6=	*Eptesicus capensis*	(V)	+	+
6=	*Pipistrellus nanus*	(V)	+	+
7	*Rhinolophus hildebrandtii*	(R)	+	+
8	*Miniopterus schreibersii*	(V)	+	+
9	*Tadarida ansorgei*	(M)	+	+
10	*Eptesicus melckorum*	(V)	+	+
11	*Tadarida aegyptiaca*	(M)	+	+
12=	*Rhinolophus fumigatus*	(R)	+	+
12=	*Rhinolophus landeri*	(R)	+	+
12=	*Kerivoula argentata*	(V)	+	+
13	*Tadarida chapini*	(M)	+	+
14=	*Eptesicus hottentotus*	(V)	+	+
14=	*Hipposideros commersoni*	(H)	+	
15=	*Tadarida bivittata*	(M)	+	+
15=	*Tadarida nigeriae*	(M)		+
16=	*Pipistrellus kuhlii*	(V)		+
16=	*Pipistrellus anchietai*	(V)		+
16=	*Tadarida ventralis*	(M)	+	
16=	*Tadarida lobata*	(M)		+
16=	*Nycteris thebaica*	(N)	+	
16=	*Nycteris vinsoni*	(N)	+	
16=	*Rhinolophus blasii*	(R)		+
16=	*Rhinolophus darlingi*	(R)		+
16=	*Rhinolophus simulator*	(R)		+
16=	*Rhinolophus swinnyi*	(R)		+

(Classification following Skinner and Smithers.[21] Family: H = Hipposideridae, M = Molossidae, N = Nycteridae, R = Rhinolophidae, V = Vespertilionidae)

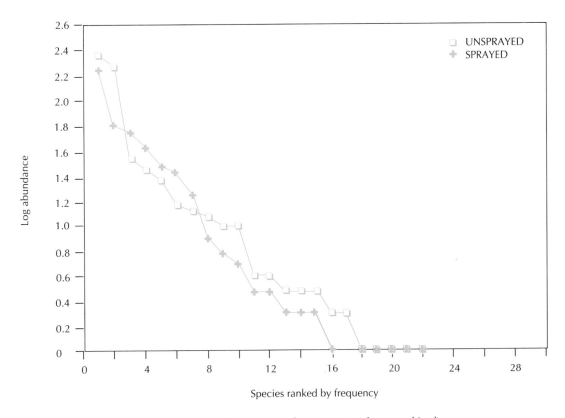

Figure 6.2 Bat species abundance in sprayed and unsprayed areas (traps and nets combined)

Box 6.1 Sampling methods for residues analysis

Bats were captured with mist nets or harp traps while foraging at night and with hand nets at tree roosts (Figure 6.4). Specimens were killed with ether after being held overnight in cloth bags and then frozen, with the exception of 9 of the 11 *Nycteris thebaica* and 3 of the 26 *Rhinolophus hildebrandtii* which were each preserved in 10% formalin in separate aluminium cans at ambient temperatures. Following Clark et al.,[22] bodies were prepared by removing the wings and feet with the skin, dissecting out the gastro-intestinal and reproductive tracts and severing the head at the base of the skull. Head musculature was dissected and placed with the body for analysis together as the carcass. Whole embryos (combining twins) from pregnant females were analysed separately, the embryonic membranes, placenta and umbilicus remaining with the carcass. The brain was extracted from the skull following removal of the top of the cranium with a scalpel. Skins, skulls, guts and reproductive tracts were preserved for taxonomic or histological analysis. Carcasses, brains and embryos were each wrapped in preweighed clean aluminium foil, weighed to 1 mg and refrozen until analysis.

Analysis for the presence of the o,p' and p,p' isomers of DDT, DDD and DDE and fat content of tissues was carried out by the Tobacco Research Station, Harare. Samples were ground with anhydrous sodium sulphate and residues extracted with hexane in a Soxhlet apparatus for 6 hours. Extracts were concentrated by evaporation and partitioned three times with acetonitrile. The lipid percentage was determined from the weight of the dried hexane fraction, whilst the three acetonitrile fractions were combined and mixed with 2% sodium chloride solution and the resulting mixture extracted twice with hexane. The hexane was then dried through anyhydrous sodium sulphate and concentrated by evaporation. Clean-up was by elution with hexane through a Florosil column (deactivated with 5% water) and residues in the eluant determined to 0.01 ppm by gas liquid chromatographs fitted with electron capture detectors. Recoveries from spiked samples exceeded 85% and results were left unadjusted. To assess the validity of comparisons with other vertebrates analysed after a different treatment at the Natural Resources Institute, duplicate samples from birds were processed by both laboratories and the residue levels proved to be in very close agreement.

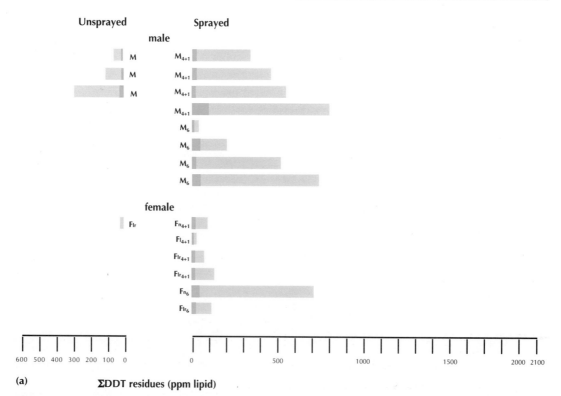

Figure 6.3 Total DDT residues in carcass at 1% fat levels for (a) *Pipistrellus rueppellii* (b) *Scotophilus borbonicus* (c) *Nycticeius schlieffeni* (d) *Eptesicus somalicus* M, male; F, female; J, Juvenile; b, non-breeding; i, immature; j, juvenile female; jr, with recent juvenile; l, lactating; lr, recently lactating; n, nulliparous; o, unsuccessful reproduction; pr, pregnant; 4, 5, 6, 4–6, number of spray treatments; +1, one year post-spray

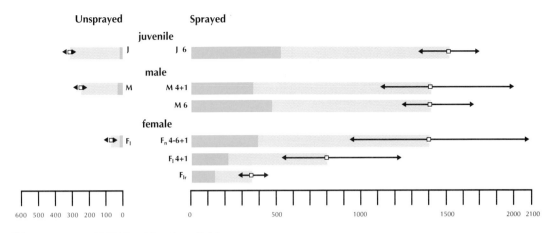

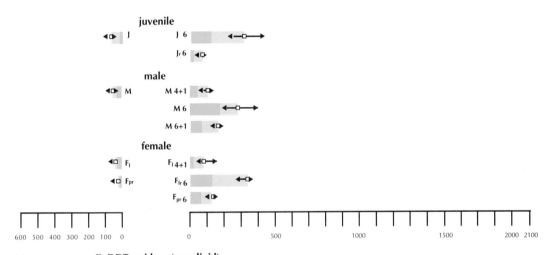

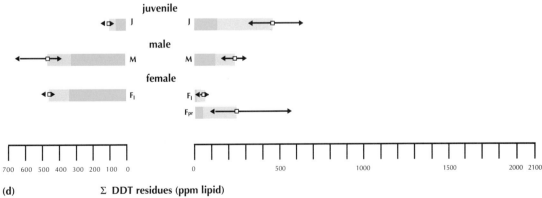

DDT in the Tropics

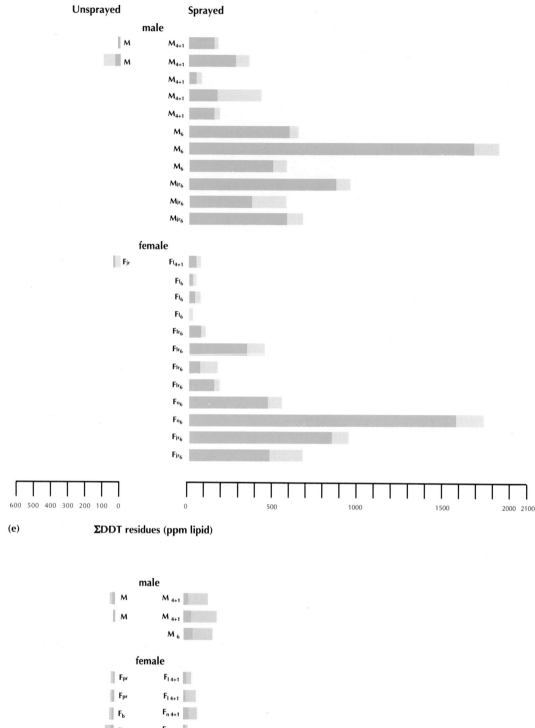

Figure 6.3 (cont.) Total DDT residues in carcass at 1% fat levels for (e) *Rhinolophus hildebrandtii* (f) *Hipposideros caffer* (g) *Tarida* spp. (h) *Nycteris thebaica*. M, male; F, female; J, Juvenile; b, non-breeding; i, immature; j, juvenile female; jr, with recent juvenile; l, lactating; lr, recently lactating; n, nulliparous; o, unsuccessful reproduction; pr, pregnant; 4, 5, 6, 4–6, number of spray treatments; +1, one year post-spray

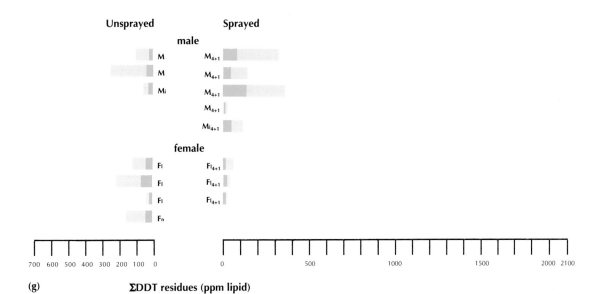

(g) ΣDDT residues (ppm lipid)

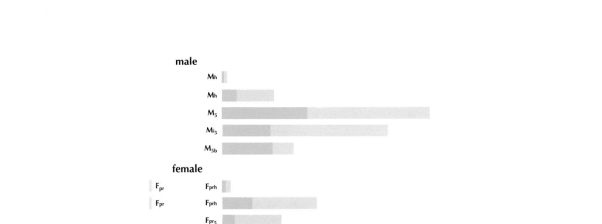

(h) ΣDDT residues (ppm lipid)

Box 6.2 Interpretation of results

Most DDD in stored wildlife samples is derived from DDT, whether preserved by freezing or in formalin, and the amount of 'original' DDT can be calculated by multiplying the concentration of DDD by 1.11 and adding this sum to the analytical value of residual DDT.[3,23,24] Accordingly, this corrected residue of DDT is distinguished from DDE and both combined to yield total DDT (ΣDDT).

Laboratory work on bats has shown that mortality is more closely related to the concentration of DDT in the brain than other tissues and, at higher concentrations, the amount of DDT in carcass lipids correlates with brain levels (Figure 6.5a).[1,25]

Thus, since DDT is stored in body lipids and bats experience pronounced fat cycles it is only when fat deposits are low that potentially lethal DDT concentrations are likely to be achieved in the brain[1] (Figure 6.5b). For the purpose of predicting potential hazard, it is necessary to extrapolate DDT concentrations in bats with seasonally varying fat levels to an empirical minimum fat content found in bats during drought or times of stress. As individuals from several species were found with fat deposits amounting to between 1–3% of body weight, a critical level of 1% body lipids will be used for calculations.

It is recognized that susceptibility to DDT and powers of detoxification will vary between species, but data obtained by Clark from the little brown bat (*Myotis lucifugus*) in the USA enables predictions of potential mortalities to be made from known DDT loads.[25] Although Clark's predicted lethal minimum level of DDT in carcass lipids was 470 ppm, to give a lethal brain load of 12 ppm wet weight, the lower 95% confidence limit of this value (360 ppm) has been conservatively used to identify bats at risk when carcass lipids have been reduced to 1% by weight. Thus, in the original study on *Myotis lucifugus*, 10 bats died out of a total of 21 bats found with >400 ppm DDT in their carcass lipids.[26]

Following conventional treatment of positively skewed data, all residue values have been log-transformed for statistical testing and geometric means provided on a wet weight basis unless otherwise specified.

(a)

Figure 6.4 Bats were caught in (a) harp traps and (b) mist nets (*Source*: B. Bettany)

(b)

The relative abundance of other nocturnal animals, bushbabies, genet cats and nightjars, was also monitored in the course of driving through sprayed and unsprayed areas at night.

DDT RESIDUES FOUND IN BATS

Although no bats were found with brain levels of DDT sufficient to cause mortality, individuals of five species retained enough residues in their bodies for lethal levels to accumulate in the brain during period of acute stress, such as a drought, on the decline of fat reserves to 1% of body weight. On such a projection, individuals of the five affected species (*Rhinolophus hildebrandtii*, *Nycteris thebaica*, *Scotophilus borbonicus*, *Hipposideros caffer* and *Eptesicus somalicus*) that were resident in the unsprayed area would never be at risk. However, potential mortalities of these species living at sites exposed to five or six annual treatments ranged from 100% to 38%, excluding adult females which lost DDT residues to juveniles during lactation (Figure 6.3).

Bats can accumulate residues via three major routes; the adult diet, direct contact from roosting and foraging sites, and from the mother's milk. There is no evidence of significant placental transfer *in utero* to the young.

Foraging Behaviour and Adult Diet

The diet of the bat species being studied was determined by detailed examination of pellets. This demonstrated both species and seasonal variation in preferred prey, to be reported elsewhere.

Nocturnal insects and their invertebrate predators contain enough DDT to be significantly concentrated in bats that feed on these potential prey in recently sprayed habitats (Box 6.4) (Table 6.2).

Different foraging habits[27–29] in the different species will alter potential exposure to DDT.

Fast-flying aerial foragers (e.g. *Tadarida* and *Pipistrellus* spp.), which take relatively uncontaminated and high-flying Diptera and Lepidoptera, will accumulate fewer residues than species such as *Nycteris thebaica* which are gleaners. Prey taken in this way from the ground or low vegetation (mostly beetles, moths, spiders and grasshoppers) will contain a higher proportion of insects 'knocked down' by spraying operations.

Foraging activity in woodland and species diversity of bats were not reduced in sprayed areas, but overall catch rates were lower at some treated locations as were the relative abundances of two particularly susceptible species, *E. somalicus* and *H. caffer* (Plate 23). The level of foraging activity strongly correlated with seasonal temperature and the relative abundance of bats was influenced by water availability, both mediated in part through the food supply.

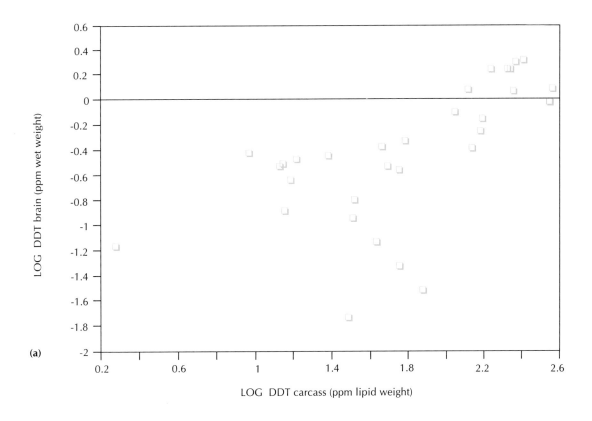

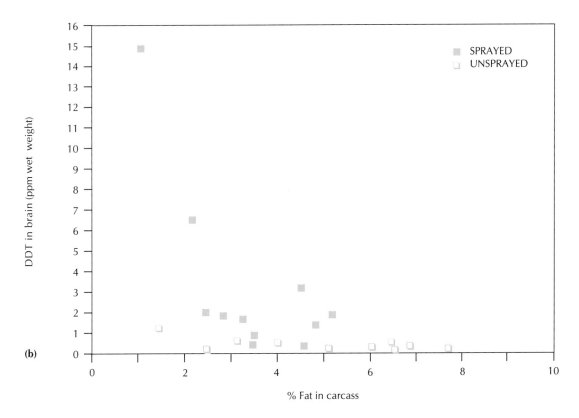

Figure 6.5 (a) Correlation between the concentration of total DDT residues in brain versus concentration in the carcass for *Rhinolophus hildebrandtii* (b) Correlation between the concentration of total DDT residues in brain versus percentage fat in the carcass.

Box 6.3 Insect sampling

Potential insect prey were sampled nightly between dusk and dawn with Malaise traps positioned along two replicate transects, simultaneously used to monitor bat activity, within both unsprayed and heavily sprayed woodland during the wet season of 1988/89 (November, December and January).

The same woodland transects were monitored during May and July 1989 to investigate insect levels during the following dry season. In addition, the insect resource base was assessed in June and July on the same nights as bats were captured around control and treated river or dam locations, where they tended to concentrate around water during the dry season.

Box 6.4 DDT residues in insects

A light trap was used to collect insects in September during the dry season from an area sprayed for the second year in succession some three months before, in June. Samples were identified and bulked by order or suborder for residue analysis (Table 6.2).

Table 6.2 DDT residues of insects (ppm wet weight)

Order	N	ΣDDT	% original DDT
Orthoptera			
Ensifera	13	2.9	57
Mantodea	5	1.7	43
Blattodea	42	1.0	42
Coleoptera	552	1.9	50
Hymenoptera	284	1.6	48
Homoptera	4859	1.7	89
Heteroptera	116	1.4	64
Diptera	4264	1.3	39
Lepidoptera (moths)	306	0.9	49

A large flightless female cockroach (Blattodea) with young, taken from a sprayed tree adjacent to the light trap, were analysed separately and found to have over double the average residues (2.6 ppm) of the more mobile trapped individuals. Butterflies (Lepidoptera) caught around dawn were also analysed separately and found to have twice the average residues (2.1 ppm) of their nocturnal relatives, the moths.

Most of the predators caught taking insects around the light trap had higher residues than their prey (wet weight): a scorpion (*Buthotus trilineatus*) = 4.0 ppm ΣDDT (60% original DDT), a gecko (*Lygodactylus* sp.) = 13.3 ppm (56% original DDT), a frog (*Ptychadena* sp.) = 3.6 ppm (75% original DDT), a toad (*Bufo* sp.) = 1.3 ppm (53% original DDT).

Lactation and Residue Concentrations in Juveniles

DDT levels were related to the reproductive state of females, being higher for pregnant or non-reproducing females than those individuals coming to the end of lactation. The data demonstrated correspondingly high levels of DDT in juveniles. Reproductively successful females can thus offload some residues each year in their milk, but at the cost of increased juvenile exposure.

In the case of *S. borbonicus*, for example, high juvenile body values of DDT could lead to very significant mortality (83%) if fat reserves fell to 1% at the end of the dry season (Table 6.3). Inexperienced young bats would be more likely to succumb to prolonged periods of reduced food supply.

Table 6.3 *S. borbonicus* at risk of gaining lethal brain concentrations of DDT (12 ppm wet weight) when carcass fat content falls to 1% of body weight, resulting from DDT concentrations of 360 ppm or more in carcass lipids.

Category	At risk	DDT Range (ppm)	N
Unsprayed area			
Juvenile	0	0–43	4
Mature Male	0	18–50	2
Lact. Female	0	7–54	4
Sprayed area			
Juvenile[6]	5	344–875	6
Mature Male[6]	3	292–610	4
Lact. Female[6]	0	111–200	3
Imm. Female[4+1]	0	292	1
Imm. Female[6+1]	1	483	1
Mature Male[4+1]	1	254–559	2
Lact. Female[4+1]	0	129–332	3

(Superscripts denote the number of annual sprays received by source areas plus any yearly interval since the last treatment)

After weaning, juveniles will also be exposed to DDT through their insect diet.

Exposure at Roosting Sites

DDT-spraying operations for tsetse control result in variable concentrations throughout the sprayed area. Operators will spray up to a height of 3–4 m along paths and water channels and selectively treat tsetse resting sites and refuges, including caves, overhangs and holes in trees caused by other animals or rot. These are also favoured roosting sites for some species of bat, and three of the five affected species are exposed to DDT in this way.

Eptesicus somalicus is thought to roost under loose bark. Both *Nycteris thebaica* and the gregarious *Rhinolophus hildebrandtii* roost in tree hollows and caves (Plates 21 and 22). When investigated, the latter species had a potential mortality of 100%, extrapolating from brain levels of DDT (19 ppm) in the captured specimens. *Nycteris thebaica* had even higher concentrations in body fat but brain levels well below the lethal limit of 12 ppm, demonstrating considerable localized detoxification in the brain. Thus, the predictive relationship between body and brain levels varies between species.

POPULATION EFFECTS

Reproduction

With one exception (Box 6.5), there was no indication of reduced reproductive success in sprayed areas as judged by the proportions of females either palpably pregnant or lactating in the breeding seasons 1989/90 and 1990/91.

Box 6.5 Reproductive failure

There was progressive evidence of reproductive failure at a colony of *R. hildebrandtii* inhabiting a baobab tree in an area sprayed for the fifth year in 1988, when observations were first made after the roost had been directly sprayed. However, the tree did not receive any further treatment although the locality was sprayed again in 1989.

During the 1988/89 breeding season, one of seven adult females failed to become pregnant at all or retain its foetus to mid/late term. Four of the remaining females lost their young. One of these bats was taken for residue analysis after it had lost its young in either late pregnancy or very early lactation. It contained the highest residue levels for an adult female, at 17.5 ppm wet weight, of which 91% would have been unaltered DDT.

The same colony of marked individuals was monitored during the 1989/90 breeding season. All five examined females were successfully pregnant in late October and a sixth escapee was probably also carrying. However, only one juvenile out of a potential five had survived through to January and 92% of its total residues, at 10 ppm wet weight, would have been unaltered DDT. One juvenile taken in November while only half-grown had already accumulated residues to the same level and percentage of unchanged DDT; easily enough to cause death by poisoning on low fat reserves.

There seems little doubt that the high juvenile mortality of this colony is directly attributable to uptake of DDT via their mother's milk, through maternal grooming after contact with DDT sprayed in the roost, as well as insect food.

Sex Ratio

For all species, differences in the sex ratios of either juveniles or adults, expressed as the number of females per male, between the unsprayed and treated areas did not reach significant levels over the study period. However, the proportion of mature males of *Eptesicus somalicus* in the sprayed area was about half that of the unsprayed area. This trend is in the predicted direction of differential mortality as a result of adult females being able to off-load DDT residues every year during lactation.

Age Structure

There were no significant differences in age structure, as judged by tooth wear between the populations in sprayed and unsprayed areas, except in the case of *Rhinolophus hildebrandtii* where juvenile mortality in one colony within a sprayed area was very high (Box 6.5).

Body Condition

In general, there is a seasonal body weight cycle, with highest fat levels after the rains and the lowest levels, and therefore the point of maximum potential morbidity, at the end of the cold dry season.

However, there was no general difference between sprayed and unsprayed areas in the condition of the three most commonly captured species, *Nycticeius schlieffenii*, *Eptesicus somalicus* and *Scotophilus borbonicus*, as indicated by body weight.

Relative Abundance

Availability of Insect Prey

Insects were sampled to see how their abundance changed with spray treatment, habitat and season.

There were no consistent differences in numbers, biomass or relative composition of insect catches between unsprayed and sprayed areas and on occasion there was also significant variation between replicates within either unsprayed or treated areas (Figures 6.6–6.9). Although considerable effort was made to reduce the effects of environmental variability by selecting comparable sites, flying insects are a very patchy resource and it can only be concluded that the effects of DDT spraying operations can't be distinguished from background environmental variation. However, this does imply that insect life, as indicated by the broad parameters under consideration, has not been severely disturbed by several annual DDT applications.

Seasonal and habitat factors relating to the presence of water appeared to have an overriding influence on the availability of insects, also found to be largely reflected in the relative abundance of bats (Boxes 6.6 and 6.7) (Figure 6.10).

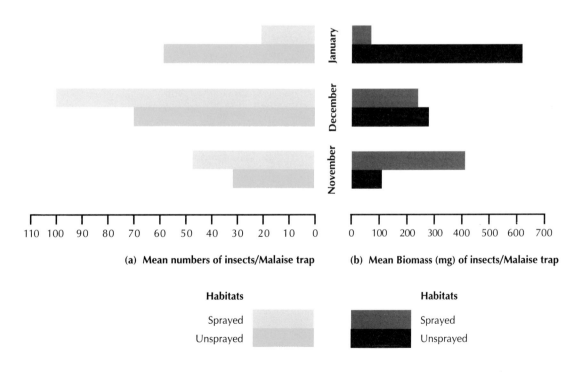

Figure 6.6 (a) Mean numbers and (b) mean biomass of insects per Malaise trap during November 1988–January 1989

DDT in the Tropics

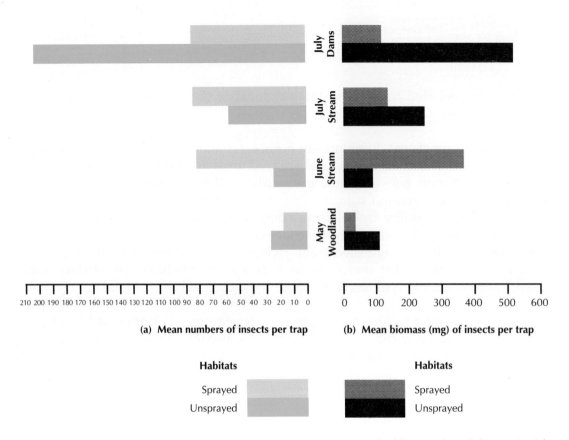

Figure 6.7 (a) Mean numbers and (b) mean biomass of insects per Malaise trap for different Siabuwa habitats, May–July 1989

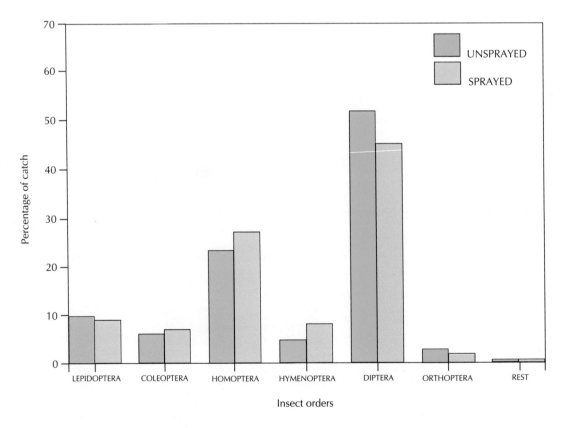

Figure 6.8 Relative composition of insects catches for sprayed and unsprayed habitats during the wet season

Nocturnal Animals

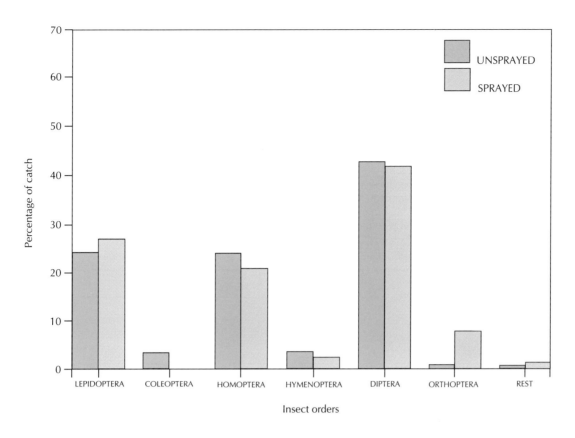

Figure 6.9 Relative composition of insect catches for sprayed and unsprayed habitats during the dry season

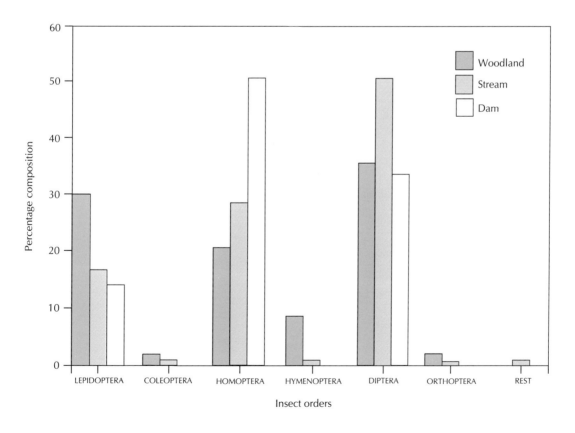

Figure 6.10 Relative composition of insect catches for different unsprayed habitats, July 1989

Box 6.6 Habitat comparisons

The patchiness of insect distribution in relation to water availability is illustrated by comparisons of catches at woodland, river and dam sites in the unsprayed area.

Wet season Both numbers and biomass of insects were significantly greater along the river bank during January, where more Homoptera and fewer Diptera were captured than in woodland.

Dry season There was a clear relationship between water availability and insect abundance during July. In addition, the relative composition of catches varied considerably between all three sites (Figure 6.10).

Table 6.4 Insect catches in different habitats during the wet and dry seasons

Wet season	River	Woodland	Probability
x Numbers/trap/night	203	59	$p < 0.05$
x Biomass (mg)/trap/night	2280	614	$p < 0.05$
Dry season	Dam	River	Woodland
x Numbers/trap/night	203	58	6
x Biomass (mg)/trap/night	523	253	48

Box 6.7 Seasonal comparisons

The same positive relationship between water and insect availability also applies to overall wet and dry season comparisons in woodland or on seasonal rivers in the unsprayed area, as already partly indicated above for selected months.

Table 6.5 Overall seasonal comparisons

Woodland	Wet	Dry	Probability
x Numbers/trap/night	53	14	$p < 0.001$
x Biomass (mg)/trap/night	281	76	$p < 0.01$
River	Wet	Dry	Probability
x Numbers/trap/night	203	45	$p < 0.001$
x Biomass (mg)/trap/night	2280	183	$p < 0.001$

BAT NUMBERS

Since sampling was carried out at focal sites around water in the middle of the sprayed and unsprayed blocks, there is unlikely to have been significant population exchange. This is supported by the marked difference in residue levels between the populations in the sprayed and unsprayed areas (Figure 6.3). The only exception to this lay in the high levels of unaltered DDT residues in *Eptesicus somalicus* lactating females in November 1989, captured in the unsprayed area, suggesting recent contamination.

Further capture-recapture figures for two of the most common species indicate no significant overlap between populations of the sprayed and unsprayed areas.[27,28,29]

There were no consistant differences between comparable catches from sprayed and unsprayed areas in either the catch rate of adults or their proportions of the total catch. The differences were greater between different habitats.

The catch-rates of bats caught in harp traps, the bulk of the fauna, were lower on the sprayed section of the Songu river (adjacent to a DDT mixing site) and at pools in sprayed woodland in both dry and wet seasons, although not significantly so on the river during the wet season. *E. somalicus*, a very small bat susceptible to DDT poisoning with readily depleted fat reserves, was the second most common bat in the unsprayed area, comprising 32% of the fauna. However, it was relatively less common at various times in all three habitat categories in the sprayed area, where overall it contributed only 12% to the sprayed fauna. This may partly be accounted for by movement into the unsprayed area during dry spells but does not explain significant declines during the rains, a time of peak activity. It seems likely that this species has suffered a general decline in the sprayed area, also mirrored by its larger congeners *E. capensis* and *E. melckorum*.

However, catch rates of the netted fauna, which largely excluded *Eptesicus*, were similar around pools in unsprayed and sprayed woodland during peak activity in the rains (Figures 6.11—6.13). Indeed numbers were marginally greater on the sprayed Songu in the dry season, although numbers were generally low.

There was no indication of generally reduced bat numbers around permanent water at the dam in the most heavily sprayed zone, where catch rates of both traps and nets were in fact higher than at the larger unsprayed dam — probably due to the former acting as a more concentrated focus of activity (Figure 6.11). However, DDT-susceptible species in the trapped fauna, *E. somalicus* and *H. caffer*, were significantly less common around the treated area which had been used as a mixing site for DDT spraying operations.

There were large increases in the catch rates of bats during the wet season, reflecting an increase in insect availability induced by the rains.

Although the potential for substantial DDT induced mortality exists among several species of bats that carry high DDT loads in times of stress, this has had little effect at the population and community level over the study, encompassing a period that did not include a major drought, and the sprayed area was not disadvantaged in terms of species richness. However, the exceptions were *E. somalicus*, a very small bat with generally low fat reserves, and *H. caffer*, some individuals of which had inadequate reserves to survive even a night of capture and a species that appeared delicate and susceptible to environmental stress.

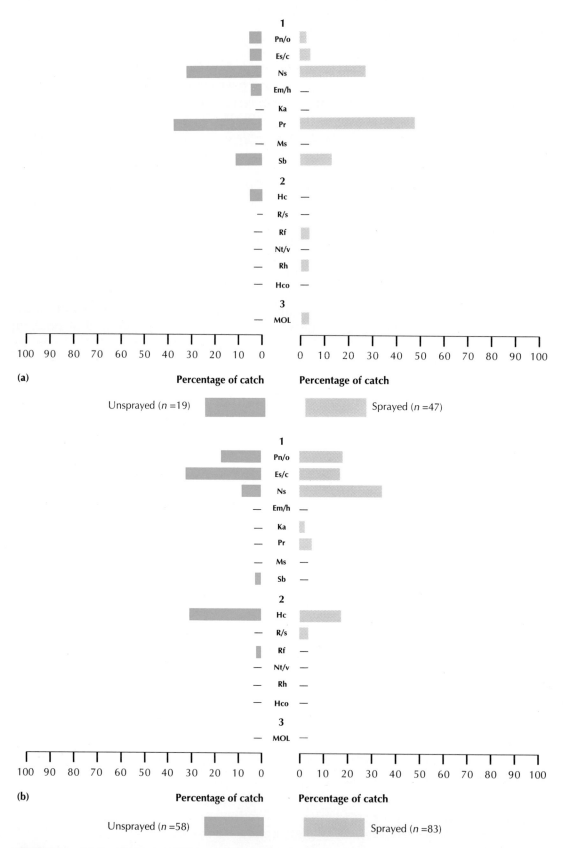

Figure 6.11 Relative abundance of bats captured by (a) nets and (b) traps at dam sites in the dry season 1 = Aerial insectivores, canopy level or below (Vespertilionidae), 2 = Foliage gleaners (Hipposideridae, Rhinolophidae, Nycteridae), 3 = Aerial insectivores, above canopy (Molossidae).
Species: Pn/o - *Pipistrellus nanus*, *P.anchietai* and *P.kuhlii*; Es/c = *Eptesicus somalicus* and *E.capensis*; Ns = *Nycticeius schlieffenii*; Em/h = *Eptesicus melckorum* and *E.hottentotus*; Ka = *Kerivoula argentata*; Pr = *Pipistrellus rueppelli*; Ms = *Miniopterus schreibersii*; Sb = *Scotophilus borbonicus*; Hc = *Hipposideros caffer*; R/s = *Rhinolophus simulator*, *R.swinnyi*, *R.blasii*, *R.darlingi*, and *R.landeri*; Rf = *Rhinolophus fumigatus*; Nt/v = *Nycteris thebaica* and *N.vinsoni*; Rh = *Rhinolophus hildebrandtii*; Hco = *Hipposideros commersoni*; MOL = *Tadarida* spp.

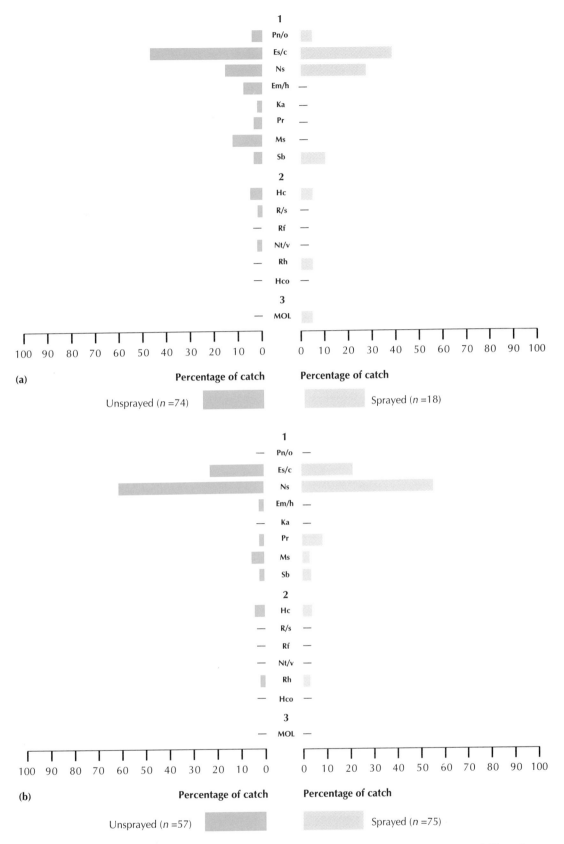

Figure 6.12 Relative abundance of bats captured in the (a) dry season (b) wet season at river sites 1 = Aerial insectivores, canopy level or below (Vespertilionidae), 2 = Foliage gleaners (Hipposideridae, Rhinolophidae, Nycteridae), 3 = Aerial insectivores, above canopy (Molossidae).
Species: Pn/o - *Pipistrellus nanus, P.anchietai* and *P.kuhlii*; Es/c = *Eptesicus somalicus* and *E.capensis*; Ns = *Nycticeius schlieffenii*; Em/h = *Eptesicus melckorum* and *E.hottentotus*; Ka = *Kerivoula argentata*; Pr = *Pipistrellus rueppelli*; Ms = *Miniopterus schreibersii*; Sb = *Scotophilus borbonicus*; Hc = *Hipposideros caffer*; R/s = *Rhinolophus simulator, R.swinnyi, R.blasii, R.darlingi,* and *R.landeri*; Rf = *Rhinolophus fumigatus*; Nt/v = *Nycteris thebaica* and *N.vinsoni*; Rh = *Rhinolophus hildebrandtii*; Hco = *Hipposideros commersoni*; MOL = *Tadarida* spp.

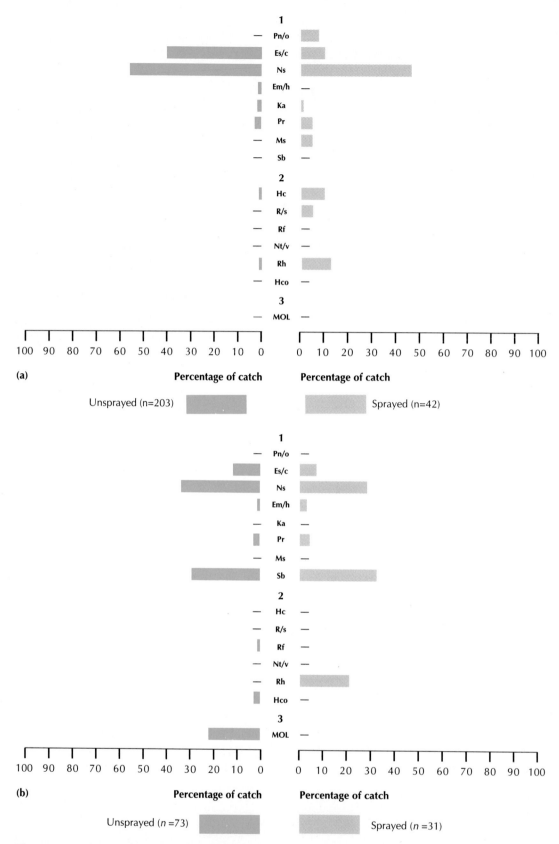

Figure 6.13 Relative abundance of bats captured by (a) traps (b) nets in the wet season at pool sites 1 = Aerial insectivores, canopy level or below (Vespertilionidae), 2 = Foliage gleaners (Hipposideridae, Rhinolophidae, Nycteridae), 3 = Aerial insectivores, above canopy (Molossidae).
Species: Pn/o - *Pipistrellus nanus, P.anchietai* and *P.kuhlii*; Es/c = *Eptesicus somalicus* and *E.capensis*; Ns = *Nycticeius schlieffenii*; Em/h = *Eptesicus melckorum* and *E.hottentotus*; Ka = *Kerivoula argentata*; Pr = *Pipistrellus rueppelli*; Ms = *Miniopterus schreibersii*; Sb = *Scotophilus borbonicus*; Hc = *Hipposideros caffer*; R/s = *Rhinolophus simulator, R.swinnyi, R.blasii, R.darlingi,* and *R.landeri*; Rf = *Rhinolophus fumigatus*; Nt/v = *Nycteris thebaica* and *N.vinsoni*; Rh = *Rhinolophus hildebrandtii*; Hco = *Hipposideros commersoni*; MOL = *Tadarida* spp.

Box 6.8 Trapping Programme

During the course of assessing bat activity on transects through woodland with the QMC bat detector, it became apparent that sites with water attracted bats for foraging and drinking from the surrounding habitats where they had also proved difficult to capture in any numbers.

It was decided to systematically sample comparable sites in unsprayed and sprayed areas with mistnets and harp-traps (Figure 6.4) from three main habitat categories, varying in their water availability. These were permanent water at Nabusenga and Chaba dams in the unsprayed and most heavily sprayed areas respectively, unsprayed and sprayed sections of the Songu river, small pools on woodland streams or free-standing pools in woodland (Plates 8 and 9).

Sites in these categories were sampled for two to four nights in succession on 13 monthly field surveys between June 1989 and January 1992, with only a night separating unsprayed from sprayed surveys. To further avoid any confounding lunar or temporal effects, sampling was undertaken in the dark phase of the lunar cycle and, although traps were open from dusk to dawn, nets were visited every 15 minutes for three hours from 15–20 minutes after sunset, to cover the main evening foraging period.

Although catch effort was designed to be equally distributed between unsprayed and sprayed sites on each monthly assessment, there were inevitable discrepancies due to trap or net damage in situ — taken into account in the comparison between categories. However, the overall effort was even — Dry season Trap nights: Unsprayed (US) = 112, Sprayed (SP) = 111. Wet season Trap nights: US = 87, SP = 83. Dry season Net nights: US = 75, SP = 75. Wet season Net nights: US = 63, SP = 54.

Before addressing separate seasonal comparisons of trapped and netted fauna between the different unsprayed and sprayed habitats, the equal catch effort enables the unsprayed and sprayed faunas to be compared in their totality. In all, thirty taxa were recognized of which 18 were common to both areas, 4 species found only in the unsprayed area and 8 recorded solely from the sprayed zone (Table 6.1). Little significance can be attributed to the latter 12 area-specific records, since 10 of these are represented by only one individual, with some rare species awaiting museum confirmation. Thus, although there was a higher species diversity in the sprayed area (Figure 6.1), the evenness of the pattern of distribution of individuals between species was similar — both areas with a long tail of species represented by only one record (Figure 6.2).

Bat Activity

Neither the activity or broad composition of the bat fauna were affected by DDT application in woodland, although the monitoring technique is biased towards more common species, some of which were not susceptible to DDT poisoning (Box 6.9).

The climate imposes a markedly seasonal pattern on bat activity, relating to temperature and insect availability, already shown by the increased catch rates of bats during the wet season.

Box 6.9 Bat activity in woodland

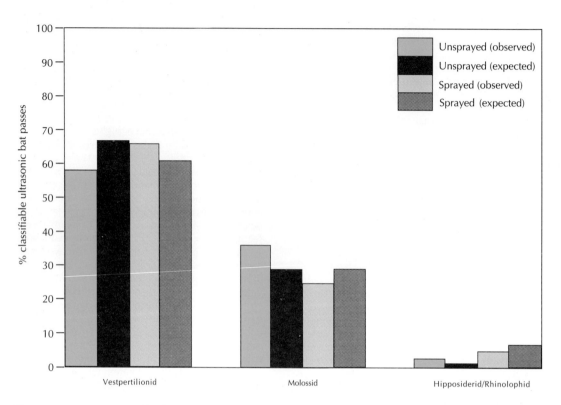

Figure 6.14 Composition of bat fauna detected in sprayed and unsprayed habitats 1988–1989

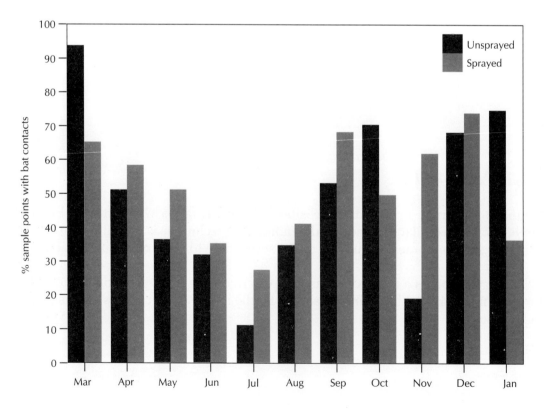

Figure 6.15 Bat activity in woodland for sprayed and unsprayed habitats

The activity of the insectivorous bat fauna was assessed along replicate transects on tracks through both untreated and sprayed woodland every month between March 1988 and January 1989, before initiation of the systematic capture program and sampling for residue analysis. Both transects in the sprayed area had received four annual DDT treatments up until July and, following spraying operations, five treatments from August through to January.

The number of bat passes were monitored from the roof of an landrover for five minutes with a QMC bat detector set at 40 kHz in 16 spot locations at intervals of ten minutes every 100 m along each track. Replicate transects within each area were assessed on subsequent nights with at most a day between unsprayed and sprayed zones. Simultaneous measurements of temperature, relative humidity and wind speed were taken at each sample point and monitoring was started 15 to 20 minutes after sunset, to cover the early evening peak in foraging activity.

The bat detector would not have recorded the very low intensity sounds of Nycterid bats, whereas high-energy audible vocalisations of Molossid bats and constant frequency signals of Hipposiderid or Rhinolophid bats were readily distinguishable (mainly *Rhinolophus hildebrandtii*, since other species would barely come into range). The remaining vocalisations are grouped together as Vespertilionid, the bulk of bat species. However, echolocation calls of the Emballonurid, *Taphozous mauritianus*, seen only once and never captured, could have been grouped either with the Molossids or Vespertilionids. Nevertheless, any mis-classifications would have been common to both unsprayed and sprayed areas and in fact there was no significant difference in the broad composition of the bat fauna (Figure 6.14), within the limitations imposed by the sampling method. Points of interest are that Molossid bats were clearly more common than indicated by capture statistics, as they generally forage at or above canopy level and the slightly greater frequency of Hipposiderid or Rhinolophid (probably *R. hildebrandtii*) bats in the sprayed zone again reflects the better roost availability of this area (more rock shelters).

Similarly, there was no consistent difference between unsprayed and sprayed areas in the general level of bat activity, as revealed by the proportion of sample points with bat contacts (Figure 6.15), with variability between replicates often as high as between control and treated zones. During November, the higher activity in the sprayed area was related to greater insect availability and biomass on the same transects, possibly accounted for by previously greater local rainfall. In contrast, both insect and bat activity were depressed in the sprayed area during January due to heavy rain.

However, there was a pronounced seasonal trend with low activity in the coldest months of the dry season, between June and August, when insect abundance was also at its nadir. Thus, there was a strong correlation between the proportion of sample points with bat contacts and mean transect temperature ($r_s = 0.46$, $n = 43$, $p = 0.002$), but a lack of any overall correlation with either relative humidity or wind speed.

OTHER NOCTURNAL INSECTIVORES AND CARNIVORES

Box 6.10 Methods

Surveys of nocturnal animals were carried out with a hand held spotlight from a Landrover along standard transects in unsprayed and sprayed areas, varying in length from 9 to 14 km along a main road and 2 to 4 km along tracks in woodland. Transects were monitored every month between May and December 1988 and between October 1989 and January 1990, with unsprayed and sprayed areas mostly surveyed between three and four hours after sunset on the same night or if not, at the same time but only one or two nights apart.

Although a wide variety of wildlife was seen during the course of about 1000 km of transects, only three groups were seen sufficiently often to merit analysis. These were bushbabies (almost certainly all the lesser bush baby, *Galago moholi*), genet cats (probably all the large-spotted genet, *Genetta tigrina*) and night jars (all species treated together).

Bushbabies

The average sighting frequency for bushbabies in vegetation along the main road was 0.20 animals per kilometre, and there was no significant differences between unsprayed and treated sections in the number of animals encountered (Figure 6.16; Plate 24), although sighting frequency was in fact slightly greater in the sprayed zone. A lower average sighting frequency of 0.18 animals per kilometre was recorded along tracks, which had poorer visibility, through similar woodland — again with no significant difference between treated and control areas.

However, there were noticeable seasonal trends related to the extent of vegetation cover and climate. Thus, sighting frequency was highest in October, averaging 0.46 animals per kilometre, when vegetation cover (ground and tree) was least dense and temperatures very high. This contrasted with the lowest sighting frequency in June, averaging 0.1 animals per km, when grass cover and leaf cover was still present but temperatures very low.

Three specimens were taken for residue analysis from around the dam that was also used as a DDT mixing site in the heavily treated area. These animals, consisting of an old male together with both an old and young adult female, had been feeding predominantly on beetles towards the end of the dry season and all contained less than 0.05 ppm ΣDDT in either their carcasses or brains. Gum exudate from trees is also reported to be a major dietary constituent.[21]

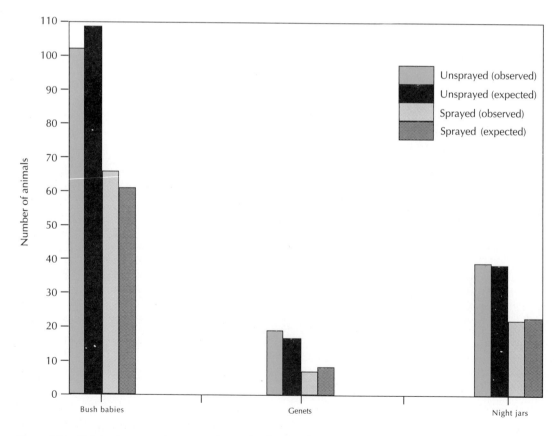

Figure 6.16 Sighting frequency for nocturnal animals other than bats in sprayed and unsprayed areas

Genet cats

These animals, also being tree climbers and taking a wide range of prey, including insects, small mammals, birds, reptiles and amphibians might be expected to accumulate pesticides. However, there wasn't any reduction in their sighting frequency in the sprayed area (Figure 6.16), although overall numbers were low.

Nightjars

Sightings of these insectivorous birds were also equally distributed between control and treated areas (Figure 6.16), indicating an absence of any deleterious effect at the population level from DDT spraying.

Conclusions

It is concluded, from the absence of any decline in the relative abundance of bushbabies or genet cats in the sprayed area that these species, potentially exposed to DDT both through diet and tree-climbing habit, are not at risk from spraying operations. Neither did road counts indicate that nightjars, as a group, were threatened in the sprayed zone although it is recognised that some of these species are migrants and a more detailed investigation of populations of resident species is required.

CONCLUSIONS

There were strong relationships between the degree of exposure to DDT and residue levels, with the exception of *E. somalicus*, apparently a local migrant, such that bats from control areas were never at risk from DDT poisoning and individuals of some species (*R. hildebrandtii*, *N. thebaica*, *S. borbonicus*, *H. caffer*) taken from areas treated annually for several years retained potentially lethal levels. However, residues declined during the course of natural detoxification in those bats taken a year post-spray.

The concentration of DDT residues in *R. hildebrandtii* and *N. thebaica* is likely to be a combination of direct exposure from sprayed roost surfaces and uptake through insect prey. Both these species and congeners, together with *Hipposideros* spp., which hunt for prey around vegetation close to the ground are at greater risk than aerial insectivores by being more likely to select non-target arthropods killed or affected by DDT spraying. Thus, there was an increased consumption of such insects by small mammals in areas subject to aerial spraying of insecticide.[30]

One problem with predicting risk of poisoning from carcass residues is the likely variation between species in their ability to detoxify or excrete DDT. Thus, there was a significant elimination of residues in urine among some shrews.[31] An adaptive enlargement of the liver was manifest in pigeons that allowed an uptake of DDT which would have killed cowbirds in a few days, a species that experienced no significant liver enlargement.[32] Similar factors might explain the relative lack of DDT accumulation by the most common species, *N. schlieffenii*, although a more detailed analysis of diet is planned.

Embryos of all species analysed in this study appeared to be efficiently protected by the placenta from accumulation of residues, acquiring only a fraction of their mother's pesticide load. Similarly, Clark and Lamont[33] found that recently born young had only 8.7% of the amount of maternal DDE although residue levels of mothers and young have also been correlated in other studies.[22,34] However, because of their low fat levels they are at greater risk if their mothers have high body burdens; exposure as neonates to lower residues than those detected in this study were responsible for later hyperactivity among adult mice.[9] Low lipid levels in foetuses, 2.2% compared to 16.9% in mothers, were also found in free-tailed bats.[35] Nevertheless, females of all species in the sprayed area, even those with substantial DDT loads, appeared able to bring their young successfully to term.

A major redistribution of pesticide residues occurs with the start of lactation, at a stage where obtaining evidence of juvenile mortality is difficult unless access to roost sites is possible, as in the case of a colony of *R. hildebrandtii*. Here, juvenile mortality was high and potentially lethal DDT residues were clearly transferred to young in milk, reducing the risk of poisoning among lactating females. Although the same offloading of pesticide residues was apparent in other species with young at risk of poisoning from high DDT levels, such as *S. borbonicus* and *H. caffer*, it was not possible to assess directly their juvenile mortality — which could have been significant. Thus, in a study of DDT poisoning of little brown bats it was concluded that juveniles were about 1.5 times more sensitive to DDT as they died with lower brain concentrations than adults.[26]

Two main periods of juvenile mortality from DDT poisoning were identified for little brown bats by Kunz *et al.*[5] The highest death rate in the pre-volant period occurred within the first 2 days of birth, while brains were small and before a lipid store could be established to buffer the young from heavily contaminated milk. In addition, evidence from other mammals suggests that organochlorines are more concentrated in milk at the beginning of lactation.[1] Further, it is known that the microsomal enzyme sytems of recently born mammals are underdeveloped, so reducing their capacity to metabolize foreign compounds.[38]

The second major period of mortality occurred around the time when young reached adult size[5], losing fat deposits as they learn to fly and forage and thereby promoting the concentration of residues in the brain. Similarly, fat depletion of young *Tadarida brasiliensis* began around first flight.[36] In the days before their hair grows, juveniles are also likely to absorb DDT dust through skin in roosts that have been directly sprayed, as significant percutaneous penetration of pesticides has been documented for young rats.[37] One possible consequence of DDT poisoning may be a poorer ability among suckling bats to retain their grip, increasing the chance of falling.[34,35]

Mature males were at greater long-term risk of DDT poisoning than females, which could annually reduce their DDT load during the course of lactation, but comparisons of sex ratio or age structure revealed no differential mortality between control and sprayed populations. Neither was there any consistent fall in the body weight of exposed bats. Thus, although the potential for significant mortality lay in the mobilization of high body loads of DDT and redistribution to the brain on the utilization of fat deposits, this was apparently not engendered by environmental conditions prevailing during the course of the study.

Although there was apparently little major impact of DDT spraying operations at the community level, it is difficult to extrapolate from impacts on individual species to the ecosystem and even to correlate residue levels in a population with a decline, since there are a host of environmental determinants that need to be taken into account. In these circumstances it is necessary to be more conservative about the use of persistent pesticides like DDT, particularly when it is impossible to monitor the fate of 'rare' species — a significant element of this bat fauna. For example, this study recorded the presence of *Nycteris vinsoni*, only the second locality record of this putative species. These bats, like their congener *N. thebaica*, are probably also very susceptible to DDT contaminated prey from a gleaning foraging style, as well subject to exposure from direct spraying of tree roosts. It is of note that during the course of the project, two reports were received from local people of the dramatic decline in bat populations following spraying of a cave and a grove of *Acacia* with numerous tree hollows.

I feel that the timing of spraying operations is perhaps optimal for the bat fauna, in that it occurs in the cold dry season when the biomass of non-target insects is least and low foraging activity indicates that many species spend time in torpor. Lower metabolic rates and reduced food intake at this time would minimize DDT uptake. Spraying operations are generally followed fairly shortly by the rains which reduce DDT concentrations in the terrestrial environment.

On the other hand, the relative abundance of bats increases in areas with greater water availability, such as along rivers which are a particular focus of spraying operations and may retain higher environmental concentrations of DDT. Although the findings of Matthiessen[20] are not strictly comparable with this and other studies, as residues were analysed in viscera and not carcasses or brains, post-rains levels of ΣDDT and the proportion of unchanged DDT were higher than those in *S. borbonicus* sampled before the rains. This could have been accounted for by accumulation of DDT in herbivorous insects over the wet season.

Residue levels were substantially reduced in bats taken from areas a year after the last spraying operation and in the only other study of this nature, DDT itself was seldom detected in bats three years after its application to Douglas-fir in north America.[4] Considering the reduced persistence of DDT in the tropics (Chapter 1), it is likely that any falls in bat populations, such as postulated for *E. somalicus* and *H. caffer*, attributable to sustained annual applications could be reversed if spraying operations ceased or were considerably reduced in scope. In any event, it is recommended that potential roost sites for bats and other animals, such as tree hollows and caves, should be treated with a short-lived insecticide that does not bioaccumulate.

REFERENCES

1. Clark, D.R., Jr. (1981) Bats and environmental contaminants: A review. *United States Department of the Interior, Fish and Wildlife Service. Special Scientific Report No. 235.* Washington DC. US Dept Interior.

2. Geluso, K.N., Altenbach, J.S. and Wilson, D.E. (1981) Organochlorine residues in young Mexican free-tailed bats from several roosts. *American Midland Naturalist*, **105**: 249–257.

3. Jefferies, D.J. (1972) Organochlorine residues in British bats and their significance. *Journal of Zoology*, **166**: 145–263.

4. Henny, C.J., Maser, C., Whitaker, J.O., Jr. and Kaiser, T.E. (1982) Organochlorine residues in bats after a forest spraying with DDT. *Northwest Science*, **56**: 329–337.

5. Kunz, T.H., Anthony, E.L.P. and Rumage, W.T. III (1977) Mortality of little brown bats following multiple pesticide applications. *Journal of Wildlife Management*, **41**(3): 476–483.

6. Esher, R.J., Wolfe, J.L. and Koch, R.B. (1980) DDT and DDE inhibition of bat brain ATPase activities. *Comparative Biochemistry and Physiology*, **65C**, 43–45.

7. Banerjee, B.D., Ramachandran, M. and Hussain, Q.Z. (1986) Sub-chronic effect of DDT on humoral immune response in mice. *Bulletin of Environmental Contamination and Toxicology*, **37**: 433–440.

8. Al-Hachim, G.M. and Fink, G.B. (1968) Effect of DDT or parathion on condition avoidance response of offspring from DDT or parathion-treated mothers. *Psychopharmacologia*, **12**: 424–427.

9. Erickkson, P., Archer, T. and Frederiksson, A. (1990) Altered behaviour in adult mice exposed to a single low dose of DDT and its fatty acid conjugate as neonates. *Brain Research*, **514**: 141–142.

10. Jefferies, D.J. (1975) The role of the thyroid in the production of sublethal effects by organochlorine insecticides and polychlorinated biphenyls. In: *Organochlorine Insecticides: Persistent Organic Pollutants*, (Moriarty, F., ed.) pp. 131–230. New York/London: Academic Press.

11. Bunyan, P.J. and Stanley, P.I. (1982) Toxic mechanisms in wildlife. *Regulatory Toxicology and Pharmacology*, **2**: 106–145.

12. Clark, D.R., Jr. and Kroll, J.C. (1977) Effects of DDE on experimentally poisoned free-tailed bats (*Tadarida brasiliensis*): lethal brain concentrations. *Journal of Toxicology and Environmental Health*, **3**: 893–901.

13. Clark, D.R., Jr., and Stafford, C.J. (1981) Effects of DDE and PCB (Aroclor 1260) on experimentally poisoned little brown bats (*Myotis lucifugus*): lethal brain concentrations. *Journal of Toxicology and Environmental Health*, **7**: 925–934.

14. Braham, H.W. and Neal, C.M. (1974) The effects of DDT on energetics of the short-tailed shrew, *Blarina brevicauda*. *Bulletin of Environmental Contamination and Toxicology*, **12**: 32–37.

15. Smithers, R.H.N. (1983) *The Mammals of the Southern African Subregion*. Pretoria: University of Pretoria.

16. Hutton, J.M. (1986) The status and distribution of bats in Zimbabwe. *Cimbebasia* (A), **8**: 219–236.

17. Fenton, M.B. (1975) Observations on the biology of some Rhodesian bats, including a key to the chiroptera of Rhodesia. *Life Science Contributions, Royal Ontario Museum*, **104**: 1–27.

18. Fenton, M.B. and Thomas, D.W. (1980) Dry season overlap in activity patterns, habitat use, and prey selection by sympatric African insectivorous bats. *Biotropica*, **12**: 81–90.

19. Fenton, M.B., Boyle, N.G.H., Harrison, T.M. and Oxley, D.J. (1977) Activity patterns, habitat use, and prey selection by some African insectivorous bats. *Biotropica*, **9**: 73–85.

20. Matthiessen, P. (1985) Contamination of Wildlife with DDT insecticide residues in relation to tsetse fly control operations in Zimbabwe. *Environmental Pollution Ser. B*, **10**: 189–211.

21. Skinner, J.D. and Smithers, R.H.N. (1990) *The Mammals of the Southern African Subregion*. Pretoria: University of Pretoria.

22. Clark, D.R., Jr., Martin, C.O. and Swineford, D.M. (1975) Organochlorine insecticide residues in the free-tailed bat (*Tadarida brasiliensis*) at Bracken Cave, Texas. *Journal of Mammalogy*, **56**: 429–443.

23. Jefferies, D.J. and Walker, C.H. (1966) Uptake of *pp'*-DDT and its post-mortem breakdown in the avian liver. *Nature*, **212**: 533–534.

24. French. M.C. and Jefferies, D.J. (1971) The preservation of biological tissue for organochlorine insecticide analysis. *Bulletin of Environmental Contamination and Toxicology*, **6**: 460–463.

25. Clark, D.R., Jr. (1981b) Death in bats from DDE, DDT or Dieldrin: Diagnosis via residues in carcass fat. *Bulletin of Environmental Contamination and Toxicology*, **26**: 367–374.

26. Clark, D.R., Jr., Kunz, T.H. and Kaiser, T.E. (1978) Insecticides applied to a nursery colony of little brown bats (*Myotis lucifugus*): lethal concentrations in brain tissues. *Journal of Mammalogy*, **59**: 84–91.

27. Barclay, R.M.R. (1985) Foraging behaviour of the African insectivorous bat, *Scotophilus leucogaster*. *Biotropica*, **17**: 65–70.

28. Fenton, M.B., Brigham, R.M., Mills, A.M. and Rautenbach, I.L. (1985) The roosting and foraging areas of *Epomophorus wahlbergi* (Pteropodidae) and *Scotophilus viridus* (Vespertilionidae) in Kruger National Park, South Africa. *Journal of Mammalogy*, **66**: 461–468.

29. Fenton, M.B. and Rautenbach, I.L. (1985) A comparison of the roosting and foraging behaviour of three species of African insectivorous bats (Rhinolophida, Vespertilionidae, and Molossida). *Canadian Journal of Zoology*, **64**: 2860–2867.

30. Stehn, R.A. (1976) Feeding response of small mammal scavengers to pesticide-killed arthropod prey. *American Midland Naturalist*, **95**: 253–256.

31. Forsyth, D.J. and Peterle, T.J. (1984) Species and age differences in accumulation of ^{36}Cl-DDT by voles and shrews in the field. *Environmental Pollution Ser. A*, **33**: 327–340.

32. Stickel, W.H., Stickel, L.F. and Coon, F.B. (1970) DDE and DDD residues correlated with mortality of experimental birds. *Pesticides Symposia*. pp. 287–294. Miami: Halos Association.

33. Clark, D.R., Jr. and Lamont, T.G. (1976) Organochlorine residues in females and nursing young of the big brown bat (*Eptesicus fuscus*). *Bulletin of Environmental Contamination and Toxicology*, **15**: 1–8.

34. Reidinger, R.F., Jr (1972) *Factors Influencing Arizona Bat Population Levels*. Ph.D Thesis. Tucson: University of Arizona.

35. Reidinger, R.F., Jr. and Cockrum, E.L. (1978) Organochlorine residues in free-tailed bats (*Tadarida brasiliensis*) at Eagle Creek Cave, Greenlee County, Arizona. In: *Proceedings 4th International Bat Research Conference* (Olembo, R.J., Castelino, J.B. and Mutere, F.A., eds.) pp. 85–96. Kenya: Kenya Literature Bureau.

36. Wilson, D.E., Geluso, K.N. and Altenbach, J.S. (1978) The ontogeny of fat deposition in *Tadarida brasiliensis*. In: *Proceedings 4th International Bat Research Conference* (Olembo, R.J., Castelino, J.B. and Mutere, F.A., eds.) pp. 15–19. Kenya: Kenya Literature Bureau.

37. Shah, P.V., Fisher, H.L., Sumler, M.R., Monroe, R.J., Chernoff, N and Hall, L.L. (1987) Comparison of the penetration of 14 pesticides through the skin of young and adult rats. *Journal of Toxicology and Environmental Health*, **21**: 353–366.

38. Conney, A.H. (1967) Pharmacological implications of microsomal enzyme induction. *Pharmacological Reviews*, **19**: 317–366.

Effects on Wildlife

Aquatic Studies

7

FISH

B McCarton and W Mhlanga

INTRODUCTION

The study of the impact of DDT on fish populations was the most difficult. The practical problems of sampling fish populations in opaque environments and of processing catches were compounded by a lack of information on the biology of species found in the study area and of previous work on the ecotoxicological effects of DDT on fish populations in the tropics. Furthermore, although the study was designed with the boundaries of sprayed areas in mind, these boundaries did not constitute either the limits of DDT contamination or the boundaries to fish movement.

A field team was trained in fish sampling techniques. Methods used were gill netting, seine netting and rotenone sampling. A randomized sampling approach, stratified by spray treatment areas was adopted. Establishing an adequate sampling strategy was difficult. The headwater streams of the Sengwa river around Siabuwa were chosen as a study area because it was the focus of terrestrial studies and there were clearly demarcated sprayed areas downstream of the unsprayed area. It was also practical to travel quickly between these areas. However, sampling during the rains was problematic, due to flash flooding; in addition, the fishing season was short and species diversity low. In particular, the Tigerfish, *Hydrocynus forskahlii*, a species targetted for study was absent.[1] It was therefore decided to extend the sampling programme to Lake Kariba itself (Figure 7.1).

Sites in the estuaries of tributaries of the Ume and Sanyati rivers were selected, but the latter were abandoned due to the presence of large numbers of crocodiles, which became entangled in the nets.

Variation in the environmental factors affecting fish at the river estuaries of the Lake Kariba shoreline are considerable but an attempt was made to choose comparable sampling sites on the basis of surrounding terrain, variation in depth, slope of banks, number of dead trees and the direction of flow. Ease of access was also a consideration. The effects of DDT on fish population dynamics can only be separated from effects of other environmental factors when they are correlated repeatedly with high residue burdens.

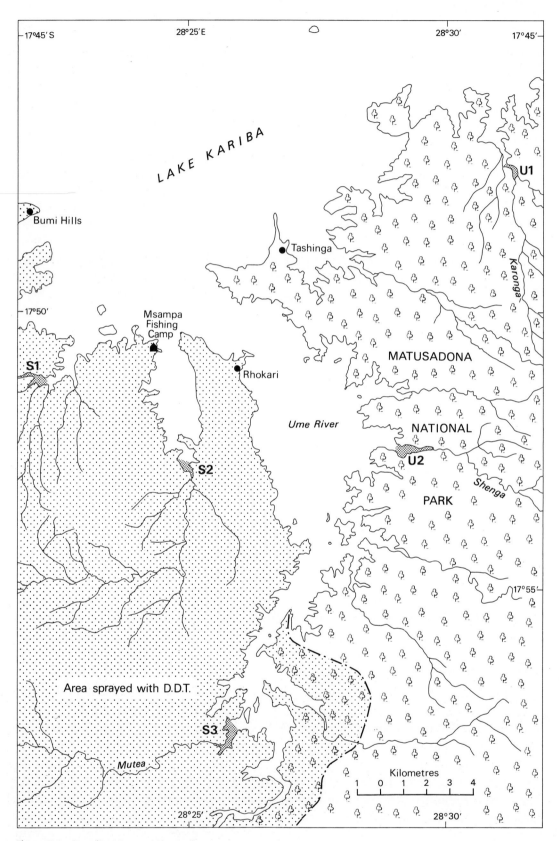

Figure 7.1 Sampling sites on Lake Kariba

THE FISH FAUNA

The Zambezi river consists of three distinct biogeographic zones: the Upper, Middle and Lower Zambezi, which are separated by the Victoria and Cabora Bassa falls, respectively. Each zone has its own fish fauna as a result of these barriers to movement and the availability of suitable habitats. Seventy-four indigenous species are known from the Upper Zambezi, 48 from the Middle and 42 from the Lower. No fish species are unique to the river and it shares many species with the River Zaire and River Nile.[2-5] The study area lies in the Middle Zambezi zone, which is now dominated by Lake Kariba.

The building of the dam at Kariba largely destroyed the natural regime in the Middle and Lower Zambezi zones. The lake began filling in 1958 and now extends over approximately 5400 km^2, to within 150 km of Victoria Falls.

Evolution of the lake ecosystem has been spectacular and is well documented.[6,7] Initially, nutrients were released, prompting high fish yields and growth of a floating water fern (*Salvinia molesta*), which covered over 20% of the lake surface in the 1960s (Figure 7.2). Waters coming into the lake, mainly from the Zambezi river, are nutrient-poor and the lake undergoes complete annual mixing and has a short retention time, resulting in progressive loss of nutrients from the system.

The introduction of the small sardine 'kapenta' (see below) was followed by a marked decline in zooplankton, a decline in the numbers of midge larvae (*Chaoborus* sp.) and an increase in Tigerfish.[8] The change from a fast-moving riverine to a slow-moving lake environment has been beneficial for some fish, such as tilapiines, which prefer the sheltered conditions. The new habitats have been colonized by nine fish species, previously found only above Victoria Falls, and six introductions have been made, giving a total of 63 species. However, not all species have benefitted from the changes.

Figure 7.2 *Salvinia* still infests the swampy margins of Lake Kariba (*Source*: B. McCarton)

Several species have disappeared from the lake and other riverine species now have more restricted distributions.[9] The fauna remains essentially riverine and further changes are likely.

The Fishery

'Kapenta' (*Limnothrissa miodon* (Boulenger, 1906)) was introduced in 1967–1968 and now supports a major pelagic fishery yielding about 20 000 tons a year. Kapenta also support large numbers of the predatory Tigerfish, (*Hydrocynus forskahlii* Cuvier, 1819) which draw anglers from all over the World, notably to the annual International Tigerfish Tournament. Overfishing inshore has affected stocks of Tigerfish which breed in rivermouths, especially the Sanyati Gorge. The average size of tigerfish caught declined from 3.0 kg in 1973, to 1.9 kg in 1980.

The inshore fishery, based on gill nets, has declined in importance since 1964 and now yields under 1000 tons a year. The catch per unit effort of Kapenta and Tigerfish has been more or less stable since 1980. International agreement with Zambia to jointly manage the fishery was reached in 1989.

Effects of DDT on Fish

DDT is acutely toxic to fish, causing unco-ordinated movements, sluggishness alternating with hyperexcitability, and difficulty in respiration, but the exact mode of action remains unclear. High DDT concentrations are lethal and laboratory tests have shown that sub-lethal levels cause tissue damage, affect nerve transmission and reduce the oxygen-carrying capacity of the blood (Box 7.1). Physiological changes and tissue damage may affect reproduction through decreased embryonation of eggs and increased egg mortality. Residues in egg yolk reduce the survival of fry.

The hazards of DDT to fish have been demonstrated under simplified conditions in the laboratory but the applicability of these studies to risk assessment in the field is limited. Potentially, DDT may reduce recruitment, growth and survival of fish populations. However, the chronic effects of DDT on populations in tropical freshwaters have not previously been studied and critical residue burdens associated with particular hazards are unknown. Fish species studied in temperate areas show enormous variation in their susceptibility to DDT, with Brown trout (*Salmo trutta* L., 1758) not surviving with more than 2 ppm wet weight in muscle[27], whilst *Ictalurus nebulosus* has been recorded alive with a residue burden of 2500 ppm DDD[28], cited in Edwards.[29]

Box 7.1 Toxicity of DDT and its metabolites

The acute toxicity of DDT to different fish species ranges from 0.003–0.110 ppm and sub-lethal effects have also been shown in the laboratory. DDT is absorbed through the gut and gills and dissolves in the lipid component of cell membranes, disrupting membrane function and enzyme systems.[10] Sodium retention is increased, affecting neurotransmission, and changes to the mitochondria may reduce the oxygen-carrying capacity of the blood.[12] Cell damage occurs and lesions arise in the liver, kidneys and brain,[13,14] causing progressive deterioration.[15] Physiological changes and tissue damage affect ovarian activity, reducing the proportion of embryonated eggs and increasing egg mortality,[16–19] and residues in egg yolk reduce the survival of fry, especially at the time the yolk sac is absorbed.[20–22] Intake of the insecticide can lead to reduced food consumption and conversion efficiency.[23]

DDT is more toxic than is its metabolite DDE. For example, the 96-h LC_{50} in bluegill *Lepomis macrochirus*, is 0.24 ppm for DDE, by comparison with 0.0012–0.021 ppm for DDT.[24–26]

No reports of fish kills have been associated with tsetse spraying operations in Zimbabwe and Matthiessen[30] considered the risk of kills in Lake Kariba to be low, as *p,p'* DDT concentrations in water did not exceed 0.0003 ppm (see Box 7.1). The risk to fry was considered higher as ΣDDT residue levels in eggs reached 2.2 ppm. Lake trout fry containing more than 3 ppm DDT died at the time of final absorption of the yolk sac.[20] Recruitment of many tropical freshwater fish may be stable over a wide range of fry densities but this offers no guarantee that DDT will not tip the balance, leading to reduced recruitment and population decline, especially where chronic effects on growth and survival affect the adult stock.

CONTAMINATION ROUTE AND DYNAMICS

DDT and its breakdown products may enter the aquatic environment in a variety of ways. River banks are targetted in spraying operations and some spray may drift into the water during treatment or be washed in later by rain. Similarly, pollution may arise from spray mixing camps and waste disposal pits which are invariably situated close to water sources; residues from these sites may be blown in on dust or contaminated organic matter, or be washed in during the rains.

Relatively small amounts of DDT have been found in river water entering Lake Kariba (Box 7.2). The insecticide binds to suspended particles in the water, but is broken down when the silt is removed from suspension and does not appear to accumulate in deposited silt.[30] However, residues have been shown to accumulate in the aquatic biota. Bio-accumulation occurs between the physical environment and certain invertebrates, with concentrations an order of magnitude greater in mussels, for example, than in the surrounding environment. However, there is little evidence of further bio-accumulation between invertebrates and fish.

Box 7.2 DDT residues in the aquatic system

Concentrations of ΣDDT on silt from headwater streams near Siabuwa in the wet season, ranged from n.d.–0.09 ppm dry weight ($n = 6$), similar to background levels in soil. Levels in aquatic invertebrates from the same area ranged from n.d.–8.0 ppm dry weight ($n = 5$).[31]

No water was tested for DDT residues in this study, but residue levels in samples taken by Matthiessen[30] at the mouths of 12 rivers entering Lake Kariba in 1983 were mostly below the detection limit, reaching a maximum of 8.6 ppm dry weight (or up to 0.0003 ppm). DDT and its metabolites are barely soluble in water and only p,p'DDT was found, adsorbed on suspended matter. Lower levels of p,p'DDT were found in deposited silt and concentrations of ΣDDT fell significantly over the rainy season (in a year of poor flows) from a geometric mean of 0.15 ppm dry weight (max 0.74 ppm) to 0.01 ppm (max 0.98 ppm). The proportion of unaltered DDT present remained constant (26–25%) but the proportion of DDE fell (64–54%). Matthiessen concluded that there was no annual accumulation of residues in silt.[30]

The geometric mean level of ΣDDT in samples of filter-feeding mussels (*Mutela dubia*) from 15 river mouths was 6.1 ppm lipid, with levels reaching 307–323 ppm in some individuals, indicating some bio-accumulation. The proportion of DDE found in mussels was lower than in silt (mean 22%, range 1–59%). However, ΣDDT levels in the lipids of fish and mussel were similar, suggesting that no further bioaccumulation occurs at this stage within the food chain.[1]

Results from Lake Chivero (McIlwaine) showed the same trend with ΣDDT levels (expressed on a dry weight basis) in oligochaetes, benthic insects and fish, 10 fold greater than those in plankton and silt, although the invertebrates and fish held similar residue levels.[32]

EFFECTS OF DDT ON GROWTH AND MORTALITY

To understand the significance of the accumulation of high residue burdens by fish, some quantification of the effects of uptake at various levels is required in the field situation. Individual fish may be affected, but it is more important to determine the point at which effects on populations become noticeable. As yet, few studies have looked specifically at growth, mortality and recruitment of fish in DDT-contaminated waters, and no previous work has been carried out in the tropics.

Estimations of growth and mortality are part of an analytical model for assessing the population dynamics of fish. They are determined using methods of ageing from the hard parts (e.g. scales) (Box 7.3), or by analysing a time series of length frequency distributions (Box 7.4). These techniques are complementary to each other and both have been used here.

Box 7.3 Estimating growth and mortality in fish populations by ageing fish from scales

Estimating growth Various environmental and/or physiological factors can alter growth rates in fish. Where a period of rapid growth is followed by slow growth, 'checks' are clearly visible in the growth rings of the scale (Figure 7.3a). Given knowledge about the timing of check formation, these can be used to age fish.

This method was used to age Red-eyed Labeo (*Labeo cylindricus* Peters, 1852) populations in the headwater streams at Siabuwa.

The 3rd scale of the horizontal row immediately above the joining of the dorsal end of the operculum to the head (Figure 7.3b) was taken as a 'key' scale at the time of capture. Scales were examined for growth checks under a binocular microscope and the radius from the focus to the edge of the scale and radii to checks were measured using an eyepiece graticule.

By relating scale radius to body length for a large number of fish, intermediate lengths can be estimated for each growth check on a scale.[33] Growth curves can then be plotted from the mean length at different ages (age t) for each species.

The von Bertalanffy growth expression is used as a model to describe the growth curves.[34] Growth parameters $L\check{s}$, the asymptotic length, and k, the rate at which fish attain $L\check{s}$, are estimated from the Gulland plot and the inverse Ford-Walford plot, and growth performance is assessed by calculating the growth index ϕ', which combines functions of k and $L\check{s}$.[35]

Data on the length at age t are used to calculate increments of growth throughout a fish's lifespan and separation of the increment data into different cohorts (or year classes) is then carried out. Differences between the means of growth increments of fish from the same site, but different calender years, were compared using ANOVA.[36]

Estimating mortality The total mortality coefficient (Z) for a population of fish is influenced by both natural mortality and any mortality due to fishing.[33] Where no fishing occurs, Z is equal to natural mortality and is estimated from an exponential decay curve.

One-year-old fish are often excluded from the calculation of Z, because young fish are not caught by the fishing gear with equal efficiency.

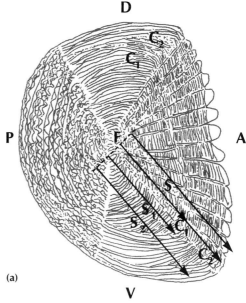

(a)

KEY

A	anterior end of scale	S_1	radius to first check
P	posterior end of scale	S_2	radius to second check
D	dorsal end of scale	S	total radius of scale
V	ventral end of scale	$C_{1,2}$	check
F	focus		

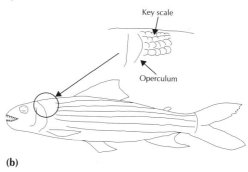

(b)

Figure 7.3 (a) Fish scale showing measurements needed for ageing determination (b) Position of the key scale from the head

The natural mortality coefficient (M) can be estimated using a formula linking $L\check{s}$ and k with mean environmental temperature.[37]

Box 7.4 Estimation of growth and mortality using ELEFAN

Electronic Length Frequency Analysis (ELEFAN) is an iterative computer programme, which provides best fit estimates of growth, mortality and recruitment parameters to selected monthly samples of length frequency data from fish catches.[38,39] This method was used on all species with adequate data sets.

Estimating growth Estimates of $L\check{s}$ and Z/k (see Box 7.3) are used to seed a series of growth curves which are then traced through the length frequency distributions sequentially arranged in time (Wetherall *et al.*, in press). A modified growth expression[37] was used in this study to take account of seasonal variation in growth. The sum of the peaks which each growth curve passes through is accumulated to identify the best curve and parameter combination.[40,41]

Estimating mortality The total mortality coefficient (Z) is calculated within ELEFAN, using the length converted catch curve. This is a plot of the natural logarithm of abundance of various age classes (N_i) of the pooled distributions in the sample against their corresponding age t. Estimation of the length at first capture (L_c) is also calculated from an approximation of the selection curve of the fishing gear.

The natural mortality coefficient (M) and fishing mortality (F) are calculated as before (Box 7.3).

GROWTH AND MORTALITY IN THE SONGU RIVER

Study sites and methods

The Songu river, rising in the project study area at Siabuwa, is typical of many tributaries flowing into Lake Kariba, its rocky bed, cut by waterfalls and scoured by flash floods, quickly drying to a few muddy pools which only persist through the dry season in wetter years. This river rises in the unsprayed area, but also flows through the area sprayed with DDT.

Two neighbouring fishing sites were selected in the area sprayed 4–5 times (S1 and S2) and two sites in the unsprayed area (U1 and U2). Hydrological measurements were made during sampling visits from 1987 to 1990 (Box 7.5). These showed that the sites were similar, with minor differences mainly attributable to their relative positions on the river.

Box 7.5 Hydrological differences between sampling sites on the Songu river

Changes in water temperature, pH, conductivity, oxygen concentration and turbidity at sampling sites were recorded throughout the study period.

Small differences between the four sites were due in part to their relative positions on the course of the Songu river, the unsprayed sites lying above the sprayed sites. Water temperatures ranged from about 17–34°C. During the cooler months of May and June water temperatures at all sites were very similar, but in the hotter months of the rains higher temperatures were recorded at the shallower unsprayed sites. Greater run-off at S1 and S2 may also have contributed to lower temperatures at these sites.

The pH of the water ranged from 6.6–9.0 and was very similar at all the sites.

Conductivity varied widely, from 40–140 µS. Conductivity fell sharply to low levels early in the rains as nutrients were flushed downstream. With the onset of drier weather conductivity began to increase again as evaporation concentrated the remaining ions and animals enriched the water with dung and urine. S1, in particular, was an important water hole for wildlife.

Oxygen concentrations also varied a good deal within the range 1–10 mg/l, tending to be slightly higher in the unsprayed area.

Turbidity was generally similar at all sites but large changes occurred over time. Turbidity at sprayed sites in January 1988 was much higher than at unsprayed sites due to the erosive power of a local rainstorm.

Sites were sampled during March, April, May and June in 1988 and 1989. Fishing was by seine netting areas approximately 120 m², blocked off by stop nets (Figure 7.4). A seine net of 10 mm stretched mesh in the centre piece and 20 mm stretched mesh in the wings, was drawn repeatedly across the site until no more fish were caught.

Species composition

Five species were caught, all five occurring in both sprayed and unsprayed sections. The Linespotted Barb (*Barbus lineomaculatus* Boulenger, 1903), Red-eye Labeo (*Labeo cylindricus* Peters, 1852) and Sharptooth Catfish, *Clarias gariepinus*, were common (Plates 25 and 26), while the Spotted Sand Catlet (*Leptoglanis rotundiceps* Hilgendorf, 1905) and Neumann's Rock Catlet (*Chiloglanis neumanni* Boulenger, 1911) were only recorded in 1989, following the heavy rains of 1988/89. As with the lake shore sites, catch per unit effort was used to compare catches (Table 7.1).

With the exception of the Sharptooth Catfish, higher catches were obtained in the sprayed area. Whether populations of these fish increased with distance downstream, or whether specific site factors were involved, is unknown. However, it is clear that any adverse effects of high DDT residue levels on numbers was secondary to natural factors.

DDT in the Tropics

Figure 7.4 Seine nets were used for fishing in shallow pools in the Songu river (*Source*: B. McCarton)

Table 7.1 Catch per unit effort (c.p.u.e) on the Songu River

	Samples	Unsprayed	Sprayed
Linespotted Barb	1988	24.4 (U1)	10.4 (S1)
	1989	23.5 (U1)	139.2 (S1)
	1989	47.6 (U2)	59.1 (S2)
Red-eye Labeo	1988	8.2 (U1)	12.9 (S1)
	1989	5.4 (U1)	17.8 (S1)
	1989	9.8 (U2)	10.5 (S2)
Sharptooth Catfish	1988	4.5 (U1)	2.7 (S1)
	1989	3.1 (U1)	2.4 (S1)
	1989	2.4 (U2)	1.0 (S2)
Spotted Sand Catlet	1989	0.1 (U1)	0.9 (S1)
	1989	0.8 (U2)	1.1 (S2)
Neumann's Rock Catlet	1989	0 (U1)	0.1 (S1)
	1989	0.5 (U2)	1.4 (S2)

DDT residue burdens

Residue levels in the Linespotted Barb, Red-eye Labeo and Sharp-toothed Catfish were significantly higher at sprayed sites than at unsprayed sites indicating populations were effectively discrete, despite the known mobility of these species (Box 7.6). The presence of high residue levels in a few individuals from the unsprayed area are suggestive of movement from the sprayed area. Highest residues were accumulated by the Sharptooth Catfish, consistent with its predatory feeding habits.

The proportion of unchanged DDT in the total residue burden was significantly different in the three species of fish caught on the Songu. Sharptooth Catfish contained the lowest proportion of DDE, and Red-eye Labeo the highest (Figure 7.5). This probably reflects the feeding habits of the different species, as predators are more likely to pick up unaltered DDT from invertebrate prey.

Fish

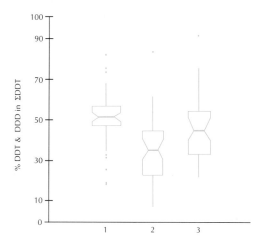

Figure 7.5 Percentage of DDT and DDD in ΣDDT in Sharptooth Catfish (1), Red-eye Labeo (2) and Linespotted Barb (3)

The proportions of different DDT metabolites in the residue burden did not vary significantly between sprayed and unsprayed sites. There was also no evidence of consistent seasonal changes in the proportion of DDE in the residue, although data were limited to 4 or 5 samples per year, taken during the dry season.

Box 7.6 DDT residues in fish from the Songu river

ΣDDT residue levels in Linespotted Barbs, Red-eye Labeo and Sharptooth Catfish from sprayed sites were significantly higher (ANOVA, $p < 0.05$, 001, 001 respectively) than levels in fish from unsprayed sites (Figure 7.6).

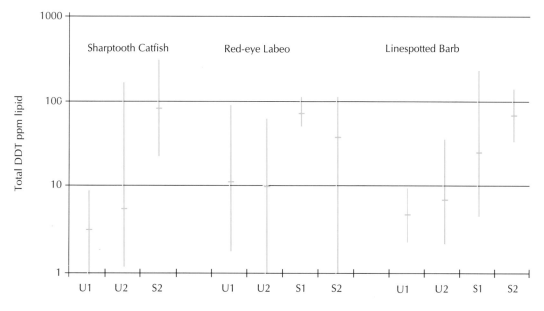

Figure 7.6 DDT residues (geometric mean and range) in Sharptooth Catfish, Red-eye Labeo and Linespotted Barb at Siabuwa

Effects of DDT on growth and mortality

Growth and mortality rates of Linespotted Barb and Red-eye Labeo varied between sprayed and unsprayed areas, although in different ways between species.

Overall growth performance in Red-eye Labeo from sprayed and unsprayed areas was similar, based on scale analysis and length-frequency data (Box 7.7). Growth increments (estimated by ageing from scales) for different cohorts of this fish in different calendar years, showed that variation from year to year outweighed differences in growth increments between sprayed and unsprayed areas.

Estimates of mortality for Red-eye Labeo based on scale analysis and length-frequency data gave different results. Mortality coefficients calculated from ageing of scales showed no differences between sprayed and unsprayed areas, whereas estimates from ELEFAN showed that in 1989 there was a difference (Box 7.8). The results show that there was great variation in mortality from year to year at both sprayed and unsprayed sites.

The growth performance of the Linespotted Barb was different at one of the sprayed sites (S2) in one year, from that at all other sites during both years of the study. The growth index in this case was higher than for any of the other sites, but as data were only available for 1989 this may represent some inherent characteristic of the site rather than an effect of DDT (Box 7.9). Mortality at this site was also higher than at other sites and this may have had an impact on growth performance of survivors. Accumulation of DDT residues was consistently greater in the fish from S2, providing circumstantial evidence that the insecticide caused the higher mortality and improved growth performance of survivors.

Growth and mortality were not examined for the Sharptooth Catfish.

Box 7.7 Growth performance of Red-eye Labeo

Based on length-frequency data There was no significant difference between the lengths attained at a given age by Red-eye Labeo from the sprayed and unsprayed areas (Table 7.2). Beyond the third year, fish from the unsprayed area appear to exhibit an increased rate of growth, evident from the increments. The mean increments between years 4 and 5, and between years 5 and 6 were 2.2 cm and 1.5 cm respectively, whereas the increment between years 3 and 4 was 1.6 cm. A progressive decline in growth increments throughout a fish's life is expected.

The von Bertalanffy growth expressions are as follows:
$L_t = 32.0 [1 - \exp 0.08 (t + 1.6)]$ UNSPRAYED AREA
$L_t = 10.3 [1 - \exp 0.82 (t + 0.2)]$ SPRAYED AREA

Despite the apparent differences in these expressions, the indices of growth performance (ϕ') for fish from sprayed and unsprayed areas were similar (Table 7.3) and expected lengths at ages 1, 2 and 3 years, calculated from the von Bertalanffy growth curves, were similar for the two treatment areas (Table 7.4).

The estimation of growth parameters using ELEFAN are given in Table 7.5. The ELEFAN program can fit a von Bertalanffy growth curve to the length frequency data using any of a combination of k and $L\check{s}$ values within the range indicated in parentheses, with the same accuracy. Table 7.5 shows that samples from the two years, 1988 and 1989 and samples from the sprayed and unsprayed areas have the same growth performance.

Thus, both methods show that growth performance of Red-eye Labeo was similar in sprayed and unsprayed sections of the Songu river.

Based on ageing from scales The ageing of fish from scales has the advantage over length frequency analysis in that growth increments of fish of known age can be followed down the years. Comparison of growth increments between and within cohorts over time may allow the timing of environmental changes to be pinpointed.

Labeo hatching in the 1984/85 wet season grew faster in their first year (reflected in 1986) than the cohorts hatching in 1985/86 and 1986/87, and in the unsprayed area the 1984/85 cohort went on to grow more in its second year than did the 1985/86 cohort (Table 7.6). The growth parameters and growth performance indices (ϕ') of two cohorts for the first three years of life in the sprayed and unsprayed areas are shown in Table 7.7. The indices confirm the better growth performance of the 1984/85 cohort over the first three years, compared with the 1985/86 cohort, but this applies equally to populations from sprayed and unsprayed areas and rules out an effect of DDT.

The most likely explanation for differences in growth performance is rainfall. Heavier falls occurred in 1984/85 than in the other two years and the difference between years was more pronounced in the northern, sprayed area, which was affected by drought in 1985/86 and 1986/87.

Table 7.2 Mean standard lengths (in cm) at ages and 95% confidence intervals for Red-eye Labeo from sprayed and unsprayed areas

Age (years)	1	2	3	4	5	6
Unsprayed area						
mean	6.1	8.0	10.1	11.7	13.9	15.4
mean + 95% C.I.	7.7	10.1	12.3	13.5	15.3	17.7
mean − 95% C.I.	4.6	5.9	7.9	9.8	12.4	13.0
n	249	89	36	21	5	2
Sprayed area						
mean	6.0	8.3	9.4	13.1	nd	14.7
mean + 95% C.I.	7.6	10.6	10.9	13.7	nd	14.7
mean − 95% C.I.	4.4	6.0	7.9	12.9	nd	14.7
n	182	48	11	2	nd	1

C.I. — confidence interval
nd — no data

Table 7.3 Parameters from the von-Bertalanffy growth expression for Red-eye Labeo

	Unsprayed	Sprayed
L_ϕ	32.0	10.3
k	0.08	0.82
t_0	−1.6	−0.2
ϕ'	1.92	1.94

Table 7.4 Expected standard lengths (in cm) of Red-eye Labeo aged 1–3 years using the growth equations with estimated parameters

Age (years)	Unsprayed	Sprayed
1	6.0	6.0
2	8.0	8.3
3	9.9	9.4

Table 7.5 Growth parameters calculated using ELEFAN for Red-eye Labeo samples taken in 1988 and 1989 from the sprayed and unsprayed areas of the Songu River

	Unsprayed			Sprayed		
	L_ϕ	k	ϕ'	L_ϕ	k	ϕ'
1988	22.0 (20.0–25.0)	0.35 (0.25–0.35)	2.23	20.0 (18.0–24.0)	0.25 (0.25–0.38)	2.00
1989	23.5 (20.0–23.5)	0.25 (0.25–0.60)	2.14	20.5 (18.0–24.0)	0.28 (0.20–0.38)	2.07

Table 7.6 Differences between the growth increments of Red-eye Labeo in different calendar years

	1st growth increment			2nd growth increment	
Unsprayed area					
Year	1988	1987	1986	1988	1987
Mean growth increment	5.71	5.78	6.13	1.58	1.63
Sprayed area					
Year	1988	1987	1986	1988	1987
Mean growth increment	5.76	5.65	6.30	1.59	1.90

NB: Underlining indicates there is no significant difference between the means.

Table 7.7 Growth parameters and growth indices for different cohorts of Red-eye Labeo from the sprayed and unsprayed areas

	Sprayed				Unsprayed			
	L_ϕ	k	t_o	ϕ'	L_ϕ	k	t_o	ϕ'
1984/85	9.74	0.81	−0.36	1.89	17.38	0.26	−1.27	1.89
1985/86	18.13	0.17	−1.39	1.75	11.42	0.38	−1.14	1.69

NB: The year of the cohort means the individuals were actually born during the wet season of that year.

Box 7.8 Effect of DDT on mortality of Red-eye Labeo

Table 7.8 Mortality estimates for Red-eye Labeo

Unsprayed							
Age (years)	0+	I+	II+	III+	IV+	V+	VI+
Number by age class		240	52	14	20	5	2
Z		1.5	1.3	−0.4	1.4	0.9	
Z (mean)							1.0
Sprayed							
Age (years)	0+	I+	II+	III+	IV+	V+	VI+
Number by age class		434	52	18	2	0	1
Z		2.1	1.1	2.2			
Z (mean)							1.8

Ageing by scales showed that few Labeo were over five years old and very few fish older than four years were caught in the sprayed area (Table 7.8). However, age-frequency regressions for fish caught in the sprayed and unsprayed areas did not differ significantly ($p > 0.05$) in slope indicating overall mortality rates in the sprayed and unsprayed areas were similar (Figure 7.7).

Mortality coefficients estimated by ELEFAN differ from those estimated from scale ageing. The mean ELEFAN estimate for the sprayed area in 1988 and 1989 was 1.26 compared with 1.8 by scale ageing, while the respective values for the unsprayed area were 2.89 and 1.0. Mortality declined significantly at unsprayed sites between 1988 and 1989 ($p < 0.05$), but remained constant at the sprayed site (Table 7.9). In neither year did mortality vary significantly between sprayed and unsprayed sites, despite an apparent difference between sprayed and unsprayed sites in 1988. Non-significance in this case is due to the wide scatter of limited data particularly from the sprayed site.

Table 7.9 Total mortality coefficients of Red-eye Labeo from length frequency data

Site	1988	1989	Mean
Unsprayed	4.59	1.19	2.89
Sprayed	0.97	1.54	1.26

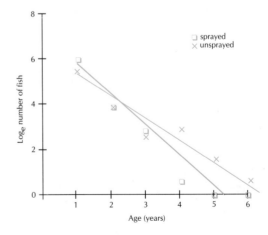

Figure 7.7 Mortality rates of Red-eye Labeo in sprayed and unsprayed areas

Box 7.9 Effect of DDT on growth performance and mortality of the Linespotted Barb

Growth performance was generally similar between years and between treatments but nested analysis of variance (NANOVA) showed that fish grew better at S2 in 1989 than at the other sites (whether they had been sprayed or not) during the same year (Table 7.10).

Table 7.10 Growth parameters estimated by ELEFAN and performance indices of Linespotted Barb

Site		1988				1989		
		L_∞	k	ϕ'		L_∞	k	ϕ'
S1	Optimum	7.5	.48	1.43	Optimum S1	8.0	.40	1.41
	Minimum	6.8	.65	1.48	Minimum	6.5	.60	1.40
	Maximum	10.0	.28	1.44	Maximum	9.6	.30	1.44
					Optimum S2	8.0	.75	1.68
					Minimum	7.0	1.10	1.73
					Maximum	13.0	.30	1.71
U1	Optimum	9.3	.35	1.48	Optimum U1	8.8	.35	1.43
	Minimum	7.0	.65	1.50	Minimum	7.0	.63	1.49
	Maximum	10.8	.25	1.46	Maximum	10.5	.28	1.48
					Optimum U2	9.0	.30	1.39
					Minimum	7.0	.45	1.34
					Maximum	11.5	.20	1.42

NB: S1 and S2 were sprayed for the 4th time in 1988 and 5th in 1989
U1 and U2 were not sprayed

Table 7.11 Total mortality coefficients (Z) of Linespotted Barb from sprayed and unsprayed sites

Site	1988	1989
S1 — sprayed 4–5 times	2.11	2.65
S2 — sprayed 4–5 times		4.02
U1 — unsprayed	2.37	2.44
U2 — unsprayed		1.60

Natural variation in mortality rates was high for this species, but the only significant differences occurred between sites within treatment areas in 1989 and between U2 and S2 ($p < 0.01$) (Table 7.11).

The fish at S2 had the highest residue burdens and the improvement of growth performance at this site may have resulted from reduced competition for resources due to DDT-induced mortality. However, there was no evidence that resources limited growth at other sites.

GROWTH AND MORTALITY IN LAKE KARIBA

Study sites and methods

Study sites were selected in the estuaries of seasonal rivers entering Lake Kariba, to monitor the effects of DDT coming from the catchment areas.

Two sites were selected in the area sprayed with DDT in 1988 and 1989 (S1 and S2), and one site at the mouth of the Mutea river, the headwaters of which were sprayed in 1984, 1988 and 1989 (S3) (Figure 7.1). Catches from these sites were then compared with catches from two matched sites in the unsprayed area of Matusadona National Park: one at the mouth of the Karonga River (U1) and the other in the mouth of the Shenga River (U2). These sites were chosen for their similarity in terms of ease of access; number of dead, but standing trees in the estuary; surrounding terrain; apparent depth; and slope of banks (Table 7.12).

Although the local history of spraying operations was known there was no independent evidence to show that contamination varied between sites or that sites were ecologically similar in other respects. Hydrological measurements were taken to confirm that all the sites were broadly similar. Measurements included water temperature, pH, conductivity, oxygen concentration and turbidity. Two sites (U2 and S3) were affected by run-off from seasonal rivers to a greater extent than sites on the lake shore. This resulted in higher conductivity; lower oxygen concentration (in the case of S3); lower pH (particularly at S3, possibly due to the prevalence of *Salvinia molesta*); and higher nutrient concentrations. The up-river sites also showed higher zooplankton counts, probably due to dispersal of plankton by the wind at the more exposed sites on the lake shore.

Table 7.12 Some features of the sampling sites important in their influence on fish populations

Feature	Site				
	U1	U2	S1	S2	S3
DDT treatment	None	None	1988, 1989	1988, 1989	1984, 1988, 1989
Locality	Lake shore	Up-river	Lake shore	Lake shore (lagoon)	Up-river
Slope	Shallow	Steep	Shallow	Shallow	Steep
Commercial fishing activity	Some gill netting	Sport	Some gill netting	Gill netting	Sport

Sampling was carried out using a fleet of gill nets of mesh sizes 38–127 mm stretched mesh, designed to trap the larger species of fish present. Nets were set between 1600 and 1800 h and lifted at 0600 h the following day. Fishing was carried out twice per month around the new moon. The sampling order of the sites was randomized so that the same sites were not fished at the same time each month.

Sampling began in May 1988 and continued until February 1990. Two months of sampling were possible before ground spraying occurred adjacent to S3 and three months before spraying adjacent to sites S1 and S2.

Individual fish were measured (total and standard lengths), weighed, sexed and their reproductive state determined.

Species composition

Seventeen of the 63 fish species inhabiting Lake Kariba were caught in the gill netting programme (Table 7.13).

Table 7.13 Fish caught in gill nets on Lake Kariba

Zambezi Parrotfish	— *Hippopotamyrus discorhynchus* (Peters, 1852)
Bulldog	— *Marcusenius macrolepidotus macrolepidotus* (Peters, 1852)
Cornish Jack	— *Mormyrops (Mormyrops) deliciosus* (Leach, 1818)
Eastern Bottlenose	— *Mormyrus longirostris* Peters, 1852
Imberi	— *Brycinus imberi* (Peters, 1852)
Tigerfish	— *Hydrocynus forskahlii* Cuvier, 1819
Manyame Labeo	— *Labeo altivelis* Peters 1852
Purple Labeo	— *Labeo congoro* Peters 1852
Butter Catfish	— *Schilbe (Schilbe) mystus depressirostris* (Peters 1852)
Sharptooth Catfish	— *Clarias gariepinus* (Burchell, 1822)
Vundu	— *Heterobranchus longifilis* Cuvier & Valenciennes, 1840
Brown Squeaker	— *Synodontis zambezensis* Peters, 1852
Mozambique/Kariba Tilapia	— *Oreochromis (Oreochromis) mossambicus/mortimeri* (species not segregated)
Green Happy	— *Serranochromis (Sargochromis) codringtonii* (Boulenger, 1908)
Purpleface Largemouth	— *S. (S.) macrocephalus* (Boulenger, 1899)
Northern Redbreast Tilapia	— *Tilapia rendalli rendalli* Boulenger, 1896
Banded Tilapia	— *T. sparrmanii* A. Smith, 1840

All species were common to all sites, except the Banded Tilapia, which was restricted to S1 and infrequently caught. For quantitative differences to be detected in species composition it is necessary to take account of fishing effort, as this may differ between sites. Hence catch per unit effort (c.p.u.e.) as measured in numbers of fish per net day (a net day represents one net set for a 12-h period) gives an appropriate unit for making assessments.

C.p.u.e. varied significantly with season so comparisons before and after spraying are restricted to the same monthly periods (Table 7.14). In most months differences in c.p.u.e. between sites were less than the seasonal differences occurring within sites.

Table 7.14 Catch per unit effort by month and site before and after the first and second spray treatments

	Site	Month				
		May	Jun	Jul	Aug	Sep
Before first treatment	S1	9.4	14.6	7.4	5.2	
	S2	9.4	5.6	8.8	10.2	
	S3				4.2	3.3
	U1				12.2	
	U2				2.8	
After first treatment	S1	12.2	18.1	22.5	14.3	10.7
	S2	5.2	4.2	5.5	6.0	4.6
	S3	27.5	18.1	10.0	7.6	6.1
	U1	10.9	8.6	37.3	7.9	20.4
	U2	29.6	22.1	14.7	17.6	25.2
		Oct	Nov	Dec	Jan	Feb
Before second treatment	S1	2.8	3.8	19.0	10.1	
	S2	7.1	3.1	7.0	37.2	
	S3	4.0	4.6	11.3	34.6	
	U1	6.0	44.2	43.1	30.8	
	U2	8.1	5.6	20.6	25.1	
After second treatment	S1	16.9				26.9
	S2	4.8				16.4
	S3	13.0				27.7
	U1	47.4				32.2
	U2	19.4				51.3

There were no significant differences comparing c.p.u.e. within sites before and after spray treatments, between sites within treatment areas or between treatment areas within spray treatments (nested ANOVA, $p > 0.05$ in all comparisons).

Catch composition showed some changes following spraying, particularly for Manyame Labeo and for Kariba/Mozabique Tilapia. Natural variation was high, but some effect of the insecticide in reducing catch per unit effort for these species cannot be discounted.

The composition of the catch in terms of fish families was broadly similar between the unsprayed sites and those sprayed twice, and the changes following spraying were also comparable (Box 7.10). Composition of the catch at the site sprayed three times (S3) was quite different, but similar to the unsprayed site situated up the Ume estuary (U1). The changes in composition following spraying were also consistent with changes at the unsprayed site and thus DDT spraying does not appear to affect catch composition at the family level.

Box 7.10 Catch composition at sprayed and unsprayed sites on the lake shore

There is no evidence DDT affected faunal composition at the family level. Figures 7.8 and 7.9 show the percentages of the major families of fish caught at different sampling sites before and after spraying.

Several differences in catch composition between sites can be attributed to habitat differences rather than effects of DDT. A major factor influencing catch composition at the up-river sites was the breeding behaviour of potamodromous species such as Tigerfish, which inhabit these sites in large numbers at certain times of year. The decrease in Characidae (predominantly Tigerfish) at up-river sites after spraying occurred in both sprayed and unsprayed areas and was probably due to their spawning behaviour. Other seasonal factors may also be involved. For example, pre-spray fishing was carried out during the cold season, when lower water temperatures would have reduced feeding activity in many species and hence their vulnerability to capture. This may have resulted in fewer cichlids and fewer mormyrids being caught in the pre-spray catches. The relatively large increase in mormyrids at U1 after spraying may have been due to higher water temperature and increased feeding activity in the warmer months following spraying. Finally, changes in the catch composition pre- and post-spraying (for example a reduction in the percentage of catfish) were generally greater in the unsprayed area, suggesting that environmental variables contributed considerably to this change and that DDT had little or no effect.

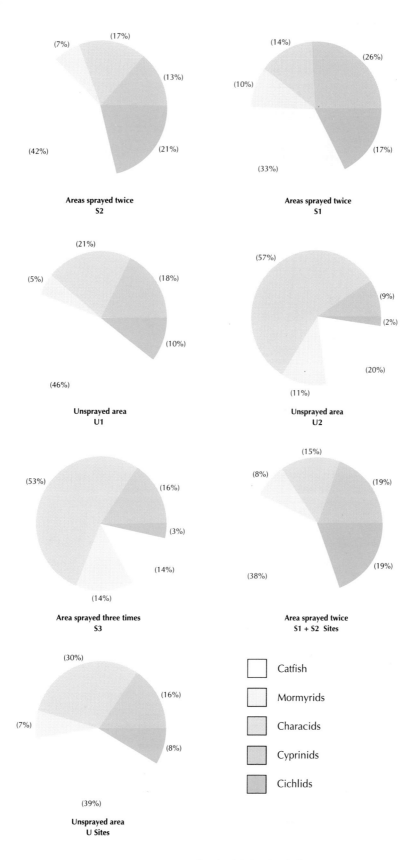

Figure 7.8 Percentage catch per unit effort for fish families before spraying

DDT in the Tropics

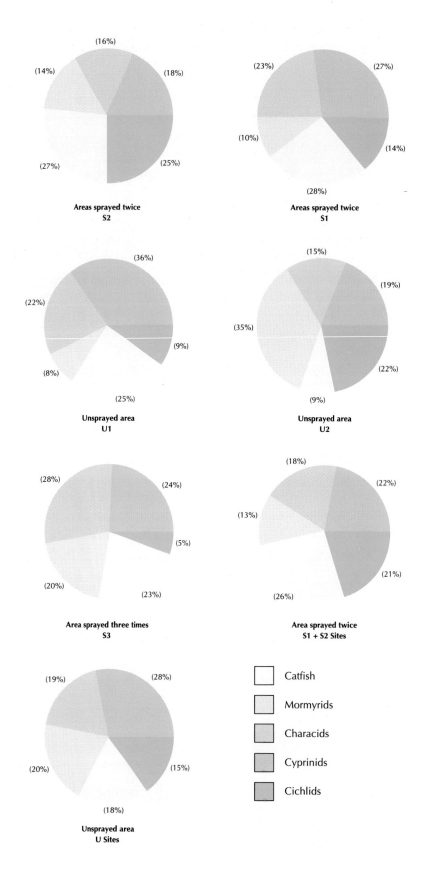

Figure 7.9 Percentage catch per unit effort for fish families after spraying

Fish

DDT residue burdens

Samples of whole fish of differing sizes caught in the gill nets at each sample site were preserved in formalin for DDT residue analyses. The residue analyses were carried out by the Natural Resources Institute.

The highest mean ΣDDT residues were found in Butter Catfish (Plate 27), but residue levels were quite similar in all species from sprayed area waters (range 1–21 ppm lipid). Highest maximum DDT levels were in Tigerfish (Box 7.11; Plate 28), followed by Sharptooth Catfish and Zambezi Parrotfish (Box 7.12).

The only fish species showing differences in residue burdens between unsprayed sites and all the sprayed sites, was the Butter Catfish (Box 7.13). This fish is omnivorous, predating small fish, snails, shrimps and other aquatic and terrestrial invertebrates, but also taking seeds and other vegetable matter. It probably moves little between sites. Its diet and sedentary habits are conducive to the accumulation of higher DDT residues in waters draining sprayed areas.

Box 7.11 ΣDDT residue in Tigerfish

Residue levels did not vary significantly between sprayed and unsprayed areas, but were significantly different between sites within treatment areas (ANOVA, $p < 0.05$). In particular, fish from the larger estuaries (U2, S3) were more heavily contaminated than those from lakeshore sites (U1, S1, S2) (Figure 7.10).

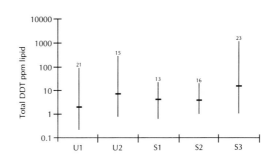

Figure 7.10 DDT residues (geometric mean, range and sample size) in Tigerfish

Box 7.12 ΣDDT residue in Zambezi Parrot Fish

Residue levels in fish from S3 were significantly higher than in fish from unsprayed sites (Student-Newman-Keuls test, $p < 0.05$) but no other significant variation was found (Figure 7.11).

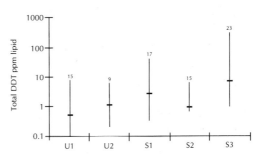

Figure 7.11 DDT residues (geometric mean, range and sample size) in Zambezi Parrotfish

Box 7.13 ΣDDT residue in catfishes

Fish from unsprayed sites were significantly less contaminated than fish from sites sprayed once (Student-Newman-Keuls test, $p < 0.05$) or from the site sprayed twice ($p < 0.01$). There was no difference in contamination levels in fish from sites sprayed once and twice (Figure 7.12).

Patterns of contamination in the Brown Squeaker and Sharptooth Catfish were similar. Fish from sites treated once were significantly more contaminated than fish from untreated sites ($p < 0.05$). However, residue levels in fish from the site treated twice were lower and did not differ significantly from levels in the other treatment areas (Figures 7.13 and 7.14).

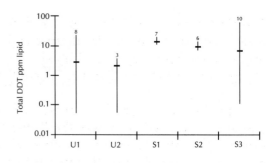

Figure 7.13 DDT residues (geometric mean, range and sample size) in Brown Squeaker

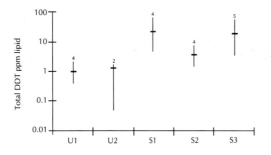

Figure 7.12 DDT residues (geometric mean, range and sample size) in Butter Catfish

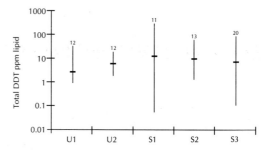

Figure 7.14 DDT residues (geometric mean, range and sample size) in Sharptooth Catfish

None of the other species showed significant differences between treatments, but all fish showed a trend towards higher residue levels at site S1 than at site S2. Both sites are mouths of rivers which drain areas sprayed at the same time in the same years and the results indicate that the amount of DDT entering the aquatic system can vary enormously and is probably dependant on the proximity of spraying to waterways, local rainfall, contamination of invertebrates and the detailed topography of the areas.

The Zambezi Parrotfish showed higher residue burdens at the site sprayed three times from those at all other sites, but no difference between the two sites sprayed twice and the unsprayed sites. This fish is sedentary, and an omnivorous bottom-feeder, taking plant matter and invertebrates. This diet is likely to lead to lower levels of DDT accumulation than those of a predator.

The other species in which DDT residue burdens were measured are more mobile, although the Brown Squeaker less so than either Tigerfish or the Sharptooth Catfish. The latter two are capable of moving great distances and spawn upstream in affluent rivers flowing into Kariba. Tigerfish showed no significant difference in residue burdens between treatment areas, but levels were different between the sites further upstream in the Ume (S3 and U2) and those on the lake-shore inlets (S1, S2 and U1).

Maximum residue levels in this species taken from the unsprayed sites were high, suggesting that DDT had been picked up elsewhere. This fish showed the highest maximum residue levels, which is consistent with its position as the fish-eating predator at the top of the aquatic food chain. Sharptooth Catfish showed similar mean residue levels to those in Brown Squeaker and they also have similar feeding habits as omnivores that will eat almost anything small enough to swallow. Brown Squeaker is a bottom feeder, whilst Sharptooth Catfish rests on the bottom, but feeds anywhere within the water column. Maximum residue levels in the latter were considerably higher, particularly at sprayed area sites. This may be explained by the high mobility of this species, bringing it into closer contact with upstream sites which are more likely to show heavier contamination from ground spraying.

The distribution of the total residue burden between DDT and its metabolites differed with species, but not from site to site. This is important, as unchanged DDT is more toxic than its metabolite DDE.[24-26] Butter Catfish had the lowest proportion of total DDT in unchanged form, and Brown Squeaker the highest (Figure 7.15). This cannot be explained on the basis of diet, as both fish have a similar range of prey, and may reflect differing abilities to degrade and metabolize DDT. However, levels varied a great deal within species.

Data for sites were pooled to examine the effect of season. None of the species showed consistent peaks reflecting influx of fresh DDT into the aquatic system at any given time of year. However, data were only available from one season for Butter Catfish and thus seasonal changes in the balance of the different DDT metabolites cannot be discounted for this species. The relatively low proportion of DDE in the residue burden of these fish is not consistent with previous findings, which reported that fish from higher trophic levels had a large proportion of their residue burden as DDE.[29]

The total DDT residue burden, however, shows changes with season, with Tigerfish showing higher residue burdens during the period following rains. Unlike findings in temperate areas, where higher residue levels in fish in the spring and early summer coincided with the period of highest agricultural use of DDT[29], there appears to be some delay between the use of the insecticide and the occurrence of high residue levels in fish in Lake Kariba.

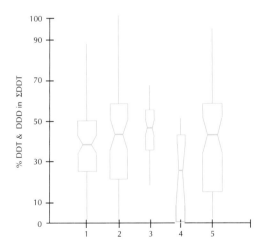

Figure 7.15 Percentage of DDT and DDE in ΣDDT in fish from Lake Kariba. 1, Tigerfish; 2, Sharptooth Catfish; 3, Brown Squeaker; 4, Butter Catfish; 5, Zambezi Parrotfish.

Effects on growth, mortality and fecundity

The Butter Catfish and Zambezi Parrotfish showed some differences in residue burdens between sprayed and unsprayed fishing sites on Lake Kariba and thus growth and mortality data for the populations of these fish were examined.

There was no evidence that growth performance of either Butter Catfish (Box 7.14) or Zambezi Parrotfish (Box 7.15) was affected by DDT spraying, but mortality appeared to be reduced after two consecutive spray treatments at one of the sprayed sites (S1) for Butter Catfish (Box 7.16) and at S3 for Zambezi Parrotfish (Box 7.17). For both species, the differences occurred at sites where the fish had higher DDT residues in their bodies and thus the lowered mortality rate may be attributed to DDT. This apparently anomolous result requires further investigation.

Box 7.14 Growth performance of Butter Catfish

The optimum, maximum and minimum values of ϕ' that correspond to the limits of the range of $L\check{s}$ and k for the Butter Catfish are shown in Table 7.15.

The growth index ϕ' values were used in order to compare growth between fish from different areas and from one spray treatment to the next. A NANOVA showed no difference in ϕ' values between sites, between treatment areas within sites or between different years spraying within treatment areas.

Table 7.15 Growth indices for Butter Catfish at sprayed and unsprayed sites over time

Site	1988		1989
	pre-spray	post-spray	post-spray
U1	2.61	2.75	2.83
U2	–	2.37	2.74
S1	2.75	2.65	2.59
S2	2.59	2.75	2.72
S3	–	2.91	2.71

Box 7.15 Growth of Zambezi Parrotfish

The growth of fish from different sites and between the treatments are compared using growth indices (ϕ'). The optimum indices in Table 7.16 reveal a small decrease in growth performance at S2 from before spraying in 1988 to after spraying in 1988. At the sprayed site S1, however, there was an increase in indices. There was no change in growth performance at the unsprayed site U1. After spraying in 1989, there was a decrease in growth performance at all sites. The decrease was most pronounced at S1, followed by that at S2. At S3 and at the unsprayed site U1 the decreases in indices were much smaller and relatively similar. A nested analysis of the variance (NANOVA) revealed that there was no significant difference in growth performance between sites within areas or among spray treatments, even though the indices showed small differences.

Table 7.16 Growth parameters estimated by ELEFAN and performance indices of Zambezi Parrotfish

Site		Before 1988 spraying			After 1988 spraying			After 1989 spraying		
		L_∞	k	ϕ'	L_∞	k	ϕ'	L_∞	k	ϕ'
U1	Optimum	34.0	.30	2.54	32.5	.33	2.54	35.5	.23	2.45
	Maximum	39.0	.27	2.61	35.0	.27	2.52	39.0	.20	2.48
	Minimum	30.0	.55	2.69	31.0	.41	2.60	31.5	.34	2.53
S1	Optimum	37.5	.23	2.50	30.0	.60	2.73	32.0	.23	2.36
	Maximum	42.0	.18	2.50	40.0	.27	2.64	40.0	.18	2.46
	Minimum	36.0	.26	2.53	28.0	.55	2.63	29.0	.40	2.53
S2	Optimum	32.0	.50	2.71	30.0	.45	2.61	32.0	.25	2.41
	Maximum	36.0	.20	2.41	33.0	.35	2.58	36.0	.20	2.41
	Minimum	28.5	.70	2.75	26.0	.70	2.68	28.0	.35	2.44
S3	Optimum				30.0	.50	2.65	34.0	.33	2.57
	Maximum				33.0	.34	2.57	40.0	.23	2.57
	Minimum				27.0	.62	2.66	29.5	.47	2.61

NB: S sites were sprayed for the first time in 1988 and again in 1989
S3 was sprayed in 1984, then again in 1988 and 1989
U1 was not sprayed

Box 7.16 Mortality of Butter Catfish

Comparison of total mortality coefficients between sites revealed differences between S3 and U2; S3 and U1; and U2 and U1. Mortality rates were higher at the sprayed site.

The mortality coefficient at S2 before spraying in 1988 was similar to those at all other sites, but once spraying commenced differences were evident. Mortality of fish from S2 was not significantly different to that at U2 throughout the study, and was similar to that at U1 before and after spraying in 1988, but different after spraying in 1989. There was a significant change in the mortality coefficient (Z) over time at S2, from before spraying in 1988 to after spraying in 1989, but no significant changes occurred at the other sites (Figure 7.16).

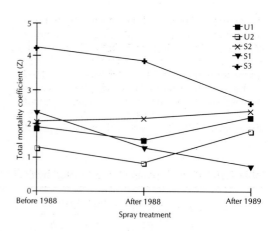

Figure 7.16 Total mortality of Butter catfish over the spray period

Box 7.17 Mortality of Zambezi Parrotfish

Figure 7.17 shows the mortality coefficients of Zambezi Parrotfish at the different sites. Mortality coefficients did not vary with time or between sites except at S3, where mortality fell after the third treatment (1989) relative to the period before. Mortality at S3 was significantly greater than at U1 or S1, and was also higher at S2 than at S1.

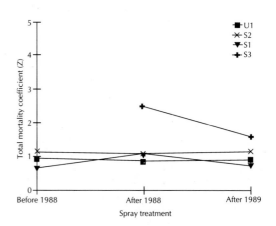

Figure 7.17 Total mortality of Zambezi Parrotfish over the spray period

The Parrotfish showed no other differences in mortality coefficients related to spray treatment, but residue burdens in this species suggest that there is no difference in the extent of contamination between the other sites. Butter Catfish showed no change in mortality coefficients at the other site that underwent two treatments (S2) or that treated three times (S3). DDT residues in fish from S2 were equivalent to those from the unsprayed sites, and thus no change in mortality coefficients over and above natural variation would be expected at this site. However, DDT burdens in fish from S3 were similar to those at S1, thus the evidence is inconclusive.

Although there was no difference in residue burdens in adult fish between unsprayed sites and those sprayed twice (S1 and S2), there was a difference in the DDT burdens in eggs. Eggs from Parrotfish taken at sites sprayed twice and three times contained enough DDT to constitute a risk to recruitment into the population (Box 7.18). Critical levels have not been determined for this fish, but the DDT burdens found far exceed the levels which kill other fish fry.

Fecundity of the Zambezi Parrotfish was also examined in relation to DDT residue burdens at different sites. Data on egg numbers in fish from sprayed and unsprayed sites is presented (Box 7.18), but is inconclusive due to the limited data set available. There is no evidence of a major impact of DDT on egg production in this fish, but minor effects cannot be ruled out.

Box 7.18 DDT residues in eggs of Zambezi Parrotfish

ΣDDT burdens in eggs from fish caught in the different treatment areas is shown in Table 7.17. Levels were higher in eggs of fish from the sprayed sites than from the unsprayed sites.

In temperate areas, the main effect of DDT on reproduction in fish was through reduced fry survival, although a reduction in egg production was associated with contamination in Brook trout[20, 21].

No work has been done on fry survival in Zambezi Parrotfish or related tropical species. However, the DDT burdens recorded here far exceed the 3.4 ppm which caused death in the Coho salmon or the 2.9 ppm which led to 100% mortality in lake trout fry[20]. There does thus appear to be a risk to recruitment into Zambezi Parrotfish populations in Lake Kariba, as a result of DDT contamination. The magnitude of this risk cannot be assessed on the basis of the data available.

The number of eggs in subsamples of the ovaries of 29 fish caught in January and February 1990 were counted. Total egg number was then estimated using the sub-sample ova number:ovary weight ratio (Table 7.18). The data suggest that there are more eggs in the ovaries of fish from the unsprayed area than the sprayed areas, but fish were generally larger in the unsprayed area.

Egg production in fish is usually dependent on size and thus observed differences may be a result of factors other than DDT.

This small data set demonstrated no clear relationship between egg number and fish length in any of the treatment areas. The results suggest that there is no major reduction in egg production in Zambezi Parrotfish in DDT sprayed areas, but further data would be necessary to establish whether more subtle changes are occurring and to assess the magnitude of the risk of the observed DDT burdens to fecundity and recruitment of this fish.

DDT in the Tropics

Table 7.17 Summary of the ΣDDT concentrations in the eggs of Zambezi Parrotfish from different treatment areas

Site	U sites (unsprayed)	S1 and S2 sites (sprayed twice)	S3 (sprayed three times)
Geometric mean	1.62	8.48	14.77
n	8	17	4
Range	0.46–5.95	4.93–14.20	6.97–23.83

The ANOVA of ΣDDT concentration in eggs of Zambezi Parrotfish from different sites revealed a significant difference in residues between sites. Further analysis (SNK) showed that the ΣDDT content of the eggs of fish from the sprayed areas was significantly greater than concentrations in eggs from unsprayed sites.

Table 7.18 Summary of fecundity analysis of Zambezi Parrotfish

	Estimated number of ova	Standard length of fish (cm)	Weight of fish (g)
Unsprayed area			
mean	6748	22.3	245
standard deviation	2781	1.5	43
range	4075–13559	19.8–25.3	162–300
n	8	8	8
Area sprayed 1–2 times			
mean	3952	18.8	131
standard deviation	2047	2.9	50
range	1210–9240	14.7–27.6	50–250
n	17	17	17
Area sprayed 2–3 times			
mean	4885	19.2	163
standard deviation	992	0.5	13
range	3595–6052	18.4–19.5	150–175
n	4	4	4

Conclusions

The results show that fish inhabiting headwater streams running through DDT sprayed areas are likely to accumulate higher residue burdens than fish in headwater streams which run through unsprayed areas or fish around the Kariba lake shore. Sedentary species on the lake shore resident in estuaries of rivers draining sprayed areas may also accumulate higher DDT than background levels. Exactly how much DDT is accumulated varies with species and is dependent both on the feeding habits of the fish and its mobility. Relatively sedentary species inhabiting waters draining from sprayed areas are likely to pick up high residue levels, particularly if they are predatory. The more mobile the fish, the more variable will be residue burdens within the population, even if the species is predatory.

There was no evidence of major adverse impacts of DDT on the fish fauna sampled and DDT did not show consistent effects on population parameters of the fish studied. Where significant effects were seen on growth and mortality, they were species-dependent. At geometric mean total DDT burdens of about 20 ppm lipid, the Butter Catfish showed no signs of being affected. However, there were signs that DDT at this level might improve survival rates of this species. The reason for this is unknown. At higher levels of accumulation (around 60 ppm) one species in headwater streams showed increased mortality and increased growth performance in survivors, whilst the other (at around 70 ppm) showed indications of reduced growth performance if environmental conditions were unfavourable and increased mortality with spraying. The implications of these findings for the population dynamics of the different species are unknown, but it is doubtful that any major changes will result at these levels of contamination.

DDT residue burdens in individual fish, particularly Tigerfish and Sharptooth Catfish (from the headwater streams) were high enough to provide a source of contamination for fish-eating birds (Plate 29).

REFERENCES

1. Matthiessen, P. (1984) *Environmental Contamination with DDT in western Zimbabwe in Relation to Tsetse Fly Control Operations*. The final report of the DDT Monitoring Project. No. K3007. London: ODA.

2. Bell-Cross, G. (1968) The distribution of fishes in Central Africa. *Fisheries Research Bulletin, Zambia*, **4**: 3–20.

3. Bell-Cross, G. (1972) The fish fauna of the Zambezi River system. *Arnoldia, Rhodesia*, **5**: 1–19.

4. Bowmaker, A.P., Jackson, P.B.N. and Jubb, R.A. (1978) *Freshwater Fishes*. pp. 1181–1230. In: *Biogeography and Ecology of Southern Africa*. (Werger, M.J.A. and Van Bruggen, A.C., eds). Den Hague: W. Junk.

5. Jackson, P.B.N. (1986) *Fish of the Zambezi system*. pp. 269–288. In: *The Ecology of River systems*. (Davies, B.R. and Walker K.F., eds) Monographiae Biologicae 60. Dordrecht, Netherlands: W. Junk.

6. Balon, E.K. and Coche, A.G. (eds) (1974) *Lake Kariba, a Man-made Tropical Ecosystem in Central Africa*. Monographiae Biologicae 24. Den Hague: W. Junk.

7. Ramberg, L., Björk-Ramberg, S., Kautsky, N. and Machena, C. (1987) Development and biological status of Lake Kariba — A man-made tropical lake. *Ambio*, **16**(6): 314–321.

8. Marshall, B.E. (1991) The Impact of the Introduced Sardine *Limnothrissa miodon* on the Ecology of Lake Kariba. *Biological Conservation*, **55**: 151–165.

9. Begg, G.W. (1974) The influence of thermal and oxygen stratification on the vertical distribution of zooplankton at the mouth of the Sanyati Gorge, Lake Kariba. *Kariba Studies*, **4**: 60–67.

10. International Programme on Chemical Safety (1989) *DDT and its Derivatives — Environmental Aspects*. Environmental Health Criteria 83. Geneva: WHO.

11. Niemi, W.D. and Webb, G.D. (1980) DDT and sodium transport in the eel electroplaque. *Pesticide Biochemistry and Physics*, **14**: 170–177.

12. Chetty, C.S., Rajendra, K., Indira, K. and Swami, K.S. (1978) *In vitro* effects of organochlorine pesticides (DDT) on catalytic potential of SDH in gastrocneius muscle of frog *Rana hedadacytyla*. *Current Science*, **47**: 842.

13. Shaffi, S.A. (1982). DDT toxicity: gluconeogenic enzymes and non-specific phosphomonoesterases in three teleosts. *Toxicology Letters*, **13**: 11–15.

14. Novikoff, A.B. (1961) In: *The Cell*. (Brachet, J. and Mirsky, A.F., eds.) New York: Academic Press.

15. McCain, B.B., Brown, D.W., Krahn, M.M., Myers, M.S., Clark, R.C. Jr., Chan, S-L. and Malins, D.C. (1988) Marine pollution problems, North American West Coast. *Aquatic Toxicology*, **11**: 143–162.

16. Crawford, R.B. and Guarino, A.M. (1985). Effects of environmental toxicants on development of a teleost embryo. *Journal of Environmental Pathology, Toxicology and Oncology*, **6**: 185–194.

17. Monod, G. (1985) Egg mortality of Lake Geneva charr (*Salvelinus alpinus*) contaminated by PCB and DDT derivatives. *Bulletin of Environmental Contamination and Toxicology*, **35**: 531–536.

18. Sukla, L. and Pandey, A.K. (1985) Effects of two commonly used insecticides on the ovarian histophysiology in the teleost *Sarotherodon mossambicus*. *Pesticides*, **19**(11): 45–48.

19. Kulshrestha, S.K. and Arora, N. (1986) Effect of carbofuran, dimethoate and DDT on early development of *Cyprinus carpio* L. Part I: egg mortality and hatching. *Journal of Environmental Biology*, **7**: 113–119.

20. Burdick, G.E., Harris, E.J., Dean, J.H., Walker, T.M., Skea, J. and Colby, D. (1964) The accumulation of DDT in lake trout and the effect on reproduction. *Transactions of the American Fisheries Society*, **93**: 127–136.

21. Macek, K.J. (1968) Reproduction in the brook trout (*Salvelinus fontinalis*) fed sublethal concentrations of DDT. *Journal of the Fisheries Research Board of Canada*, **25**: 1787–1796.

22. Broyles, R.H. and Novek, M.I. (1979) Uptake and distribution of 2,4,5,2',4',5' hexachlorobiphenyl in fry of lake trout and chinock salmon and its effects on viability. *Toxicology and Applied Pharmacology*, **50**: 299–308.

23. Pandian, T.J. and Bhaskaran, R. (1983). Food utilization in the fish *Channa striatus* exposed to sublethal concentrations of DDT and methyl parathion. *Proceedings of the Indian Academy of Sciences (Animal Sciences)*, **92**: 475–481.

24. Johnson, W.W. and Finley, M.T. (1980) *Handbook of Acute Toxicity of Chemicals to Fish and Aquatic Invertebrates*. Resource Publication 137, Washington DC: Fish and Wildlife Services, USDI.

25. Henderson, C., Pickering, Q.H. and Tarzwell, C.M. (1959) The toxicity of organic phosphorous and chlorinated hydrocarbon insecticides to fish. *Transactions of the Second Seminar on Biological Problems of Water Pollution*. Ohio: Robert A. Taft Sanitation Engineering Centre. Cincinnati: U.S. Public Health Service.

26. Marking, L.L. (1966) Evaluation of *pp'*-DDT as a reference toxicant in bioassays. *Investigations in Fish Control*, **10** Washington DC: Bureau Sport Fish and Wildlife, Fish Wildlife Service, U.S.D.I.

27. Brooks, G.T. (1974) *Chlorinated Insecticides*. Vol. II. *Biological and environmental aspects*. Parkway, Ohio: CRC Press.

28. Hunt, E.G. and Bischoff, A.I. (1960) Inimical effects on wildlife of periodic DDD applications to Clear Lake. *California Fish and Game*, **46**: 91–106.

29. Edwards, C.A. (1973) *Persistent Pesticides in the Environment*. 2nd edn. Parkway, Ohio: CRC Press.

30. Matthiessen, P. (1985) Contamination of wildlife with DDT insecticide residues in relation to tsetse fly control operations in Zimbabwe. *Environmental Pollution Ser. B*, **10**: 189–211.

31. Douthwaite, B. (1991) *Aquatic Invertebrate Populations in DDT-Contaminated and Uncontaminated waters at Siabuwa*. NRI, Chatham: DDT Impact Assessment Technical Report.

32. Greichus, Y. A., Greichus, A., Draayer, H.A. and Marshall, B. (1978) Insecticides, polychlorinated biphenyls and metals in African lake ecosystems, II Lake McIlwaine, Rhodesia. *Bulletin of Environmental Contamination and Toxicology*, **19**: 444–453.

33. Bagenal, T. (ed.) (1978) *Methods for Assessment of Fish Production in Fresh Waters*. 3rd edn. IBP Handbook No. 3. Oxford: Blackwell Scientific Publications.

34. Gulland, J.A. (1983) *Fish Stock Assessment: A manual of basic methods*. Food and Agriculture Series Vol. 1, Chichester: FAO/John Wiley.

35. Moreau, J., Bambino, C. and Pauly, D. (1986) Indices of overall growth performance of 100 Tilapia (Cichlidae) populations. pp. 201–206, In: *The First Asian fisheries forum*. (MacLean, J.L., Dizon, L.B. and Hosillos, L.V., eds.) Manila, Philippines: Asian Fisheries Society.

36. Sokal, R.R. and Rohlf, F.J. (1969) *Biometry*. San Francisco: W.H. Freeman and Co.

37. Pauly, D. (1980) The inter-relationships between natural mortality, growth parameters and mean environmental temperature in 175 fish stocks. *Journal du Conseil*. **39**: 175–192.

38. Pauly, D. and David, N. (1981) ELEFAN I, a BASIC program for the objective extraction of growth parameters from length frequency data. *Meeresforschung*, **28**: 205–211.

39. Brey, T. and Pauly, D. (1986) Electronic length frequency analysis. A revised and expanded users guide to ELEFAN 0, 1 and 2. *ICLARM Contribution*, **261**: 1–49.

40. David, N., Palomares, L. and Pauly, D. (1981) ELEFAN 0, a BASIC program for creating and editing files for use with ELEFAN I, II and III programs. Manila, Philippines: ICLARM

41. Pauly, D., David, N. and Ingles, J. (1981) ELEFAN II: Users instructions and program listing. Mimeo. pag. var.

42. Willford, W.A., Sills, J.B. and Whealdon, E.W. (1969) Chlorinated hydrocarbons in the young of Lake Michigan Coho Salmon. *Progress in Fish-Culture*, **31**: 220.

8

WATERBIRDS

R J Douthwaite

THE FISH EAGLE

DDT residue levels sufficient to cause mortality and reproductive failure have been reported in birds of prey and their eggs from Africa[32] but effects on their populations have not been studied.

In 1980 high residue levels and eggshell thinning were found in Fish Eagle *Haliaeetus vocifer* eggs collected from Lake Kariba by Ron Thomson, a Provincial Game Warden with the Department of National Parks and Wild Life Management.[1,2] Thomson's prediction[3], that the species would be extinct on the lake within 10–15 years unless the use of DDT was curtailed, caused public uproar. Surveys made between 1975 and 1987 in Matusadona National Park found no evidence of unusual breeding failure or population decline (R.D. Taylor personal communication), but residue levels in Fish Eagle eggs from the Park were relatively low[4] and the possibility that the population was declining elsewhere could not be dismissed.

The Fish Eagle is a common and conspicuous resident of the lake, and a study of the effects of DDT on its numbers was therefore seen as a priority.[5] (Box 8.1; Plate 31).

Box 8.1 Effects of DDT on the breeding Fish Eagle population of Lake Kariba

The aim of the study was to see whether variation in Fish Eagle nest densities, hatching success and eggshell thickness could be related to DDT residue levels in the egg.

Methods Three low-level aerial surveys of Lake Kariba were made during the breeding season in 1987 with the help of National Parks staff and equipment. Most of Zimbabwe's shoreline was checked, from Devil's Gorge in the south-west to Charara in the north-east, except for several islands far from shore. The location, condition and contents of every nest were recorded. Bad weather in August prevented completion of the second survey in the eastern Sanyati basin. As hatching success in this area appeared to be relatively low, three further surveys of the area were made in 1990.

Eggs ($n = 42$) were collected in 1989 and 1990 from 19 nests on Lake Kariba and one nest at Nabusenga dam, in the Siabuwa study area, using either local climbers or an Air Force helicopter fitted with a hoist (Plate 30). Egg contents were preserved in 10% formalin and later analysed at the Natural Resources Institute for lipid content and DDT residues. Eggshells were dried, measured and weighed and the Ratcliffe Index of eggshell thickness, given by the equation eggshell weight/ length x breadth[6] calculated.

Tissues from two adult birds found dead were also analysed for residues.

DDT residues and eggshell thickness Generally, clutch means of ΣDDT varied from 14–49 ppm dry weight but 113–223 ppm dry weight were found in clutches from the eastern end of the lake and the mouth of the Sengwa river (Table 8.1). Clutches from river mouths may be more heavily contaminated than those from adjacent areas if the female eats Tigerfish *Hydrocynus forskahlii* and Sharptooth catfish *Clarias gariepinus*. These species congregate in river mouths before spawning and are generally more heavily contaminated that sedentary fish, such as Brown squeaker *Synodontis zambezensis* and tilapia *Oreochromis mossambicus* (Chapter 7). All four species were recorded as prey.

Residues of DDE, DDD and DDT were found in every egg. The proportion of unaltered DDT was linked to the history of tsetse fly spraying operations nearby. DDE comprised 88–95% of the ΣDDT in clutches ($n = 11$) from areas not treated for over 6 years compared with 77–88% of the ΣDDT in clutches ($n = 9$) from areas treated in the previous year.

DDE content and the eggshell thickness were inversely related ($R^2 = 65.7\%$; F test, $p = 0.001$).

Table 8.1 Geometric mean and range of ΣDDT (ppm dry weight) levels in Fish Eagle clutches collected in 1989/90

Area	Clutches (n)	Mean	Range
Binga District	2	26	22–31
Sengwa Bay	4	45	19–120
Ume Bay	4	29	16–43
Gatshe Gatshe Communal Area	4	23	14–34
Charara and Msango bays	4	153	113–223
Redcliff Island	1	49	–
Nabusenga Dam	1	16	–

Eggshell weight, length and breadth were compared with those for eggs collected before the introduction of DDT. Comparison was restricted to eggs collected from the Zambezi river basin as significant local variation in baseline measurements of Fish Eagle eggs has been found elsewhere.[7] Four clutches collected between 1936–41 and now held in the National Museum, Bulawayo and one clutch in a private collection were measured. Length and breadth did not vary between 1936–41 and 1989–1990, but eggshell weight fell significantly ($p < 0.05$). Overall, the Ratcliffe index fell by 11% between the two periods, but eggshells from Msango Bay, at the eastern end of the Lake, were 25% thinner (Table 8.2).

Table 8.2 Variation in mean clutch eggshell thickness of Fish Eagles

	Clutches (n)	Mean Ratcliffe Index	Change on baseline (%)
Baseline, 1936–41	5	2.755	
1989–90: all	19	2.44	–11
Binga District	2	2.525	– 8
Sengwa Bay	4	2.41	–13
Ume Bay	4	2.575	– 7
Gatshe Gatshe	4	2.60	– 6
Msango Bay	3	2.07	–25

Productivity Laying on Lake Kariba occurred between late April and late July, with a peak in May. Clutch size averaged 2.20 (range 1–3, $n = 20$) and 81% of the eggs collected were viable. These figures are similar to those reported from South Africa, where clutch size averaged 2.27 eggs and 71% were viable (R.G. Davies, personal communication, 1988).

Cracked eggs (2) were found in clutches from Binga District and Msango Bay; two more eggs, apparently intact, from a nest in Charara Bay, were virtually dry inside and the shells collapsed progressively on drying.

The presence of a bird at a nest, eggs or young, constituted a breeding attempt. In 1987 hatching was confirmed at 114 nests out of 158 (72%). No laying, or hatching failure, occurred at 11 nests and was uncertain in the remaining 33 nests where adults obscured any contents. Hatching success was higher in the central section of lakeshore than at either end (Figure 8.1). Lowest success occurred in the Charara Safari Area, and particularly near Charara and Msango. In 1990, chicks were seen in fewer than half the nests here (A.N. McWilliam, personal communication, 1990).

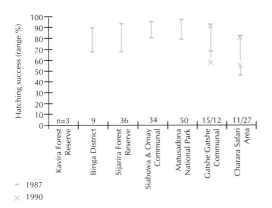

Figure 8.1 Fish eagle hatching success on Lake Kariba

The effects of DDE residues and eggshell thinning on hatching success and population size have been studied in close relatives of the Fish Eagle. In the North American Bald Eagle, *H. leucocephalus*, population decline was associated with 15–20% eggshell thinning.[8] Productivity appeared normal where DDE levels did not exceed 3 ppm wet weight, but breeding failure was almost complete when levels exceeded 15 ppm wet weight.[9] Using the same data, Nisbet[10] concluded that breeding productivity fell sharply between 2.5–5 ppm wet weight. The critical values of 2.5 and 15 wet weight equate to 17 and 100 ppm dry weight respectively. In the European White-tailed Sea Eagle, *H. albicilla*, DDE levels of 500–600 ppm lipid weight, equal to 114–136 ppm dry weight, were critical for hatching success.[11]

With geometric mean clutch residue levels of 145 ppm DDE dry weight and eggshell thinning over 20% there can be little doubt that DDT contamination contributed to low hatching success in the Charara-Msango area. A heavily contaminated clutch was also found at the Sengwa River mouth. These areas have received more spray treatments than other areas of the lake and residue levels are probably atypical of the lake generally. The shoreline in Zambia, for example, remains untreated although DDT is used by fishermen to protect their catches against post-harvest infestation.

Assuming levels ranging from 12–45 ppm DDE in the 15 clutches from other parts of the lake were typical, the risk of hatching failure can be estimated on the basis of the geometric mean and standard deviation of this sample and critical levels in the Bald Eagle. About 80% clutches would contain more than 17 ppm DDE and would be less likely to hatch because of contamination. However, no clutches would be contaminated to the extent that failure is certain (i.e. > 100 ppm).

Nest Densities Over 60% of nest sites ($n = 163$) were in dead trees, mostly killed between 1958–63 as the lake formed, and baobab trees on land. Islands and headlands are favoured nesting areas[12] where active nests were sometimes only 250–400 m apart. Major settlements on the other hand are avoided. No nests were found along 20 km of shore between Kariba town and Charara or over a similar distance near Binga. Generally, nests were closer together in areas with restricted public access, such as Game and Forest reserves, than on communal land (Table 8.3). A similar aversion to settlements is recorded in the Black Eagle *Aquila verreauxi* in Zimbabwe.[13]

Table 8.3 Spacing between occupied Fish Eagle nests along different sections of lakeshore of Lake Kariba

	Nests (n)	Median distance apart (km)	Percent > 3.5 km apart	Public access
Kavira Forest Reserve	6	1.9	0	Restricted
Binga District	21	3.0	48	Open
Sijarira Forest Reserve and Chete Safari Area	37	2.3	17	Restricted
Siabuwa and Omay Communal Areas	39	3.0	39	Open
Matusadona National Park	48	1.9	6	Restricted
Sanyati Gorge	8	2.0	0	Restricted
Gatshe Gatshe Communal Area	13	2.2	17	Open
Charara Safari Area	29	1.85	11	Open

The highest density of occupied nests was found in the Charara Safari Area, despite the presence of several fishing camps, the heaviest ΣDDT load and the lowest hatching success. A positive relationship between breeding failure and breeding frequency may explain the high density of nests in this area.

Trends in contamination Wherever possible eggs were taken from the same or adjacent territories sampled by Thomson.[4] Residue levels have increased by about 8% since 1980, rising more steeply in areas recently sprayed for tsetse fly control and falling in others. Contamination in 1980 varied with the number of previous treatments for tsetse fly control (Table 8.4) and changes since then reflect the subsequent history of spraying operations. Contamination has fallen in Binga District and Gatshe Gatshe, last treated in 1975 and 1983 respectively but, conversely, has increased in Sengwa bay, treated six times on the western shore between 1984–1989; in Ume bay, treated three times between 1988–90; and at Charara, treated twice in 1982 and 1983.

Table 8.4 Contamination (geometric means, ppm dry weight) of Fish Eagle clutches related to spray treatments of the shoreline in 1980 and 1989/90

Area	1980			1989/90			Change since 1980 (%)	Last spray
	(n)	ΣDDT	Prior treatments	(n)	ΣDDT	Sprays since 1980		
Binga	1	62	3–4	2	26	0	− 58	1975
Sengwa Bay	1	12	0	4	45	1–5	+275	1989
Ume Bay	4	21	0	4	29	1–2	+ 38	1990
Gatshe Gatshe	3	45	3	4	23	1	− 49	1983
Charara	6	88	2–5	5	122	2	+ 39	1983
Overall mean		36			39		+ 8	

The accumulation of ΣDDT residues in eggs following spraying operations may be delayed. Clutches taken from the same nest, in an area first treated in 1988 and again in 1989, contained 16 ppm in 1980, 43 ppm in 1989 and 32 ppm in 1990. Contamination in the Sengwa and Ume river mouths may therefore increase considerably before the benefits of discontinued DDT use are felt.

Hazards to adult birds Residue levels of ΣDDT in brain samples from two adult Fish Eagles found dying near Kariba town did not exceed 22.5 ppm lipid, well below a level likely to be harmful.[14,15]

Population trends, past and future Lake Kariba may now support about 550 pairs of Fish Eagles, perhaps five times more than existed on the river before the dam was built.

The breeding population in Matusadona National Park has been stable since monitoring began in 1975 (R.D. Taylor personal communication 1990) but there is little information on recent population changes elsewhere. Twenty-two pairs bred in Chete Game Reserve in 1983 (R.D. Taylor *in litt.* 1990) compared with 15 pairs in 1987. However, water levels were lower in 1987 and three sites occupied in 1983 were not surveyed in 1987 as they were thought to be too far inland. Hatching occurred at 77% of nests (17 nests, 22 chicks) in 1983 compared with 87% of nests (13 nests, 19 chicks) in 1987. The shoreline was sprayed with DDT in 1983, 1984 and 1985 but hatching success did not decline between 1983 and 1987 and the apparent reduction in breeding numbers cannot be attributed to DDE.

A claim that the population at the eastern end of Lake Kariba is declining[16] was not confirmed by local residents (K. Hustler personal communication 1988, J.D. Strydom personal communication 1990) and, away from Kariba town, the density of nesting Fish Eagles remains as high as anywhere else, despite poor hatching success and disturbance around fishing camps.

The long-term threat from DDT is now receding and levels of other pesticides in the lake are likely to remain low.[17,18] However, four birds analysed in 1986–7 from the Sanyati Basin contained very high levels of mercury (A. Renzoni personal communication 1990; K. Hustler personal communication 1988). Two of these presented clinical and post-mortem evidence of mercury poisoning. In general, the dietary concentrations of methyl mercury that impair breeding are about 1/5 those required to produce overt toxicity in adult birds of the same species.[19] The source, extent and effects of mercury pollution require further study.

Observations were too infrequent to assess fledging success accurately. Mean brood size declined significantly with age, from 1.87 at the downy stage, to 1.63 at the feathered stage (42 broods, $\chi^2 = 13.9$, $p < 0.001$), probably due to sibling rivalry[20], but the loss of complete broods near fledging was also suspected. Many nests are accessible to skilled climbers and several fishermen admitted to eating Fish Eagle chicks. Disturbance of the lakeshore by man is increasing and in the longer term the number of breeding Fish Eagles may be determined by the availability of safe nest sites. Nest sites offshore are becoming fewer each year as the dead trees collapse. As sites are lost, pairs build new nests in living mopane *Colophospermum mopane* or *Terminalia* spp., trees (R.D. Taylor personal communication 1987), often well inland and more accessible to climbers. Elephants largely destroyed the woodlands of the mid-Zambezi valley in the 1960s and 1970s, before the need to manage their numbers was recognized[21,22] and big trees now survive mainly around settlements, or where game was hunted as a means of tsetse control. Elephants continue to destroy nest sites in Matusadona National Park and in other areas and management of their numbers should not be relaxed.

In conclusion, DDT contamination is reducing hatching success on Lake Kariba but does not appear to be limiting population size. The risk of eggshell thinning should now recede, with the imminent withdrawal of DDT for tsetse fly control in Zimbabwe and a general decline in its use within the region. The number of Fish Eagles breeding on Lake Kariba is likely to fall as safe nest sites are lost and human disturbance increases.

Nineteen clutches of eggs were collected from the lake using a Zimbabwe Airforce helicopter fitted with a hoist in 1989 and local climbers in 1990. Egg contents were analysed for DDT residues and the eggshells weighed and measured. Every egg was contaminated with DDT residues and the proportion of unaltered DDT was higher in eggs from areas recently ground-sprayed for tsetse fly control. Generally, clutch means of ΣDDT varied from 14–49 ppm dry weight, but levels of 113–223 ppm dry weight were found in clutches from the eastern end of the lake and the mouth of the Sengwa River. DDE content and eggshell thickness (Ratcliffe index) were inversely related. Overall the Ratcliffe index may have fallen by about 11% since the advent of DDT, but eggshells from Msango Bay, at the eastern end of the Lake, were 25% thinner.

Low-level aerial surveys of the southern lakeshore were made in 1987 and of the Sanyati basin in 1990 to assess breeding success. In 1987, hatching success was higher in the central section of lakeshore than at either end. Breeding attempts in the Charara Safari Area, and particularly near Charara and Msango, were least successful. In 1990, chicks were seen in fewer than half the nests here (A.N. McWilliam personal communication 1990). Comparison with studies of the related North American Bald Eagle and European White-tailed Sea Eagle leave little doubt that poor hatching success in the Charara Safari Area is due to DDT residue contamination. Contamination may reduce hatching success elsewhere on the lake, but productivity generally appears high. The population of Fish Eagles appears stable and the density of nests in the Charara Safari Area is as high as anywhere else on the lake.

Comparison with Thomson's 1980 data[3,4] shows DDT residue levels in eggs have fallen in areas last sprayed several years ago, but have risen in areas recently sprayed for tsetse fly control. With the reduction in DDT use contamination should now decline generally. However, the future prospects for the Fish Eagle population appear unfavourable. Safe nest sites are being lost as dead trees in the lake collapse, and elephants destroy large trees on land. Disturbance, and the predation of chicks, by humans is already reflected in lower nest densities close to settlements than in protected reserves. This contrast will sharpen as settlement along the lakeshore increases. Mercury pollution may also be affecting adult survival and the problem requires further study.

The Reed Cormorant

DDT residue levels were measured in a large number of animal tissue samples collected between 1986–1988 by the Scientific Environmental Monitoring Group, part of European Development Fund's Regional Tsetse and Trypanosomiasis Control Programme. The largest series came from Reed Cormorants *Phalacrocorax africanus* and the data have been analysed and results published elsewhere.[17]

Samples of liver and visceral fat from 86 Reed Cormorants shot on Lake Kariba were analysed for DDT and other organochlorine insecticide residues. Unaltered DDT, or its metabolites DDD and DDE, were found in every sample. Over 97% of the ΣDDT was DDE, suggesting Reed Cormorants are remote, trophically or spatially, from fresh inputs of DDT. Nevertheless, mean residue levels were higher in the dry season than in the rains. This may be due to the presence of more lightly contaminated birds from South Africa during the rains.

The risk of increased mortality in adult birds from DDT residues is probably low. The highest residue level found in liver was 8.7 ppm ΣDDT wet weight, well below the average 185 ppm ΣDDT wet weight found in Double-crested Cormorants surviving heavy experimental exposure.[23] However, the risk of eggshell thinning and breeding failure appears considerable. Double-crested Cormorant colonies in California failed completely when residue levels in the egg reached 180–1300 ppm DDE lipid.[24] As residue levels in lipid from the egg and ovulating female should be similar[25,26] it is estimated that over 40% of female Reed Cormorants on Lake Kariba will lay eggs containing over 180 ppm DDE which will not hatch (Plate 32).

Exposure of Reed Cormorants to other organochlorine insecticides on Lake Kariba is low.

Box 8.2 DDT residue levels in Reed Cormorants: a hazard assessment

The Reed Cormorant is one of the most abundant fish-eating birds on Lake Kariba, but numbers vary through the year with more present between November and January than between March and August.[27] The opposite pattern occurs in South Africa[28,29] suggesting a single population is involved. Breeding on Lake Kariba occurs throughout the year. Food comprises mainly small cichlids, especially *Pharyngochromis darlingi* (mean weight 2.1 g; mean total length 51 mm) and *Tilapia rendalli* (mean weight 16.5 g; mean total length 88 mm), with adult males probably feeding in deeper water on predatory fish, including *Brycinus lateralis* and *Hydrocynus forskahlii*.[27]

Samples of liver and visceral fat from 86 birds were collected in the Sanyati basin, at the eastern end of Lake Kariba, between January and October 1986 and analysed for 22 organochlorine compounds.

Residues of four insecticides were detected but concentrations of three, hexachlorobenzene, HCH and lindane, were low (Table 8.5).

Table 8.5 Frequency of occurrence and maximum concentrations of organochlorine residues in Reed Cormorant tissues

Sample size	Occurrence (%)		Max. ppm lipid	
	Liver 86	Fat 58	Liver 86	Fat 58
Hexachlorobenzene	23	24	0.1	3.0
α — HCH	42	24	0.15	3.0
β — HCH	3	2	0.3	0.1
Lindane	69	52	0.2	3.0
o,p' and p,p' DDT	28	72	0.8	64.5
o,p' and p,p' DDD	57	17	0.7	12.9
o,p' and p,p' DDE	100	100	235.1	5750.0

The proportion of DDE in ΣDDT in liver varied from 97.3% in January to 99.8% in September suggesting Reed Cormorants are spatially or trophicly remote from any fresh inputs of DDT.

Residue levels in liver tended to be lower in sub-adult birds, but not significantly so. Overall geometric mean ΣDDT levels in liver were 17.5 ppm lipid (95% confidence limits 2.2–140.2) or 0.64 ppm wet weight (95% confidence limits 0.08–5.2).

ΣDDT levels in liver were significantly lower in the rains than in the dry season (Figure 8.2). There was no evidence that residue levels in fish varied seasonally and lower residue burdens in the rains are likely to reflect the presence at this time of more lightly contaminated birds from South Africa. The maximum ΣDDT residue level in Reed Cormorant livers from the southeast Cape coast was 0.25 ppm wet weight[29] compared with a mean of 0.64 ppm wet weight on Lake Kariba.

The risk of increased mortality from these levels appears to be negligible. Residues in livers of Double-crested Cormorants (*P. auritus*) surviving heavy experimental exposure to DDT, DDD and DDE averaged 40, 52 and 93 ppm wet weight respectively.[23] The highest level of ΣDDT found here was 8.7 ppm wet weight.

However, the risk of eggshell thinning and breeding failure appears considerable. DDE causes eggshell thinning in cormorants.[8,25,30,31] Adult female Reed Cormorants contained a geometric mean of 113 ppm ΣDDT lipid in their body fat ($N = 26$, 95% confidence limits, 5–2526 ppm); fat from one bird, with an embryo in the oviduct, contained 2965 ppm DDE lipid. Concentrations in lipid from the embryo and ovulating female are similar.[25,26] Complete breeding failure occurred at Double-crested Cormorant colonies in southern California where eggs contained 180–1300, arithmetic mean 671 ppm DDE.[24] By analogy with the Double-crested Cormorant, female Reed Cormorants with over 180 DDE ppm lipid in body fat are therefore likely to lay thin-shelled eggs. The sample distribution of data from Kariba indicates that over 40% of females should lay eggs containing more than 180 ppm lipid, which are unlikely to hatch.

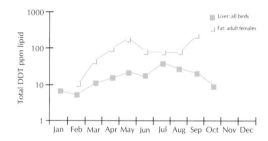

Figure 8.2 Total DDT residue levels in Reed Comorant tissues

References

1. Thomson, W.R. (1984) *DDT in Zimbabwe*. In: Proceedings of the 2nd Symposium on African Predatory Birds. (Mendelsohn, J.M. and Sapsford, C.W., eds.) pp. 169–171. Durban: Natal Bird Club.

2. Matthiessen, P. (1985) Contamination of wildlife with DDT insecticide residues in relation to tsetse fly control operations in Zimbabwe. *Environmental Pollution Ser. B*, **10**: 189–211.

3. Thomson, W.R. (1981) Letter. *Herald newspaper*, 30 April 1981, Harare.

4. Thomson, W.R. (1981) *A Report on Chemical Contamination of the Environment in Zimbabwe*. pp. 11, mimeo. Harare: Department of National Parks and Wild Life Management.

5. Douthwaite, R.J. (1992) Effects of DDT on the fish eagle *Haliaeetus vocifer* population of Lake Kariba in Zimbabwe. *Ibis*, **134**: 250–258.

6. Ratcliffe, D.A. (1967) Decrease in eggshell weight in certain birds of prey. *Nature*, **215**: 208–210.

7. Davies, R.A.G. and Randall, R.M. (1989) Historical and geographical patterns in eggshell thickness of African Fish Eagles *Haliaeetus vocifer*, in relation to pesticide use within southern Africa. In: *Raptors in the Modern World*. (Meyburg, B.-U. and Chancellor, R.D., eds.) pp. 501–513. Berlin: Weltarbeitsgruppe fur Greifvogel und Eulen e. V.

8. Anderson, D.W. and Hickey, J.J. (1972) Eggshell changes in certain North American birds. *Proceedings of the 15th International Ornithological Congress*. pp. 514–540.

9. Wiemeyer, S.N., Lamont, T.G., Bunck, C.M., Sindelar, C.R., Gramlich, F.J., Fraser, J.D. and Byrd, M.A. (1984) Organochlorine pesticide, polychlorbiphenyl, and mercury residues in Bald Eagle eggs — 1969–1979 — and their relationships to shell thinning and reproduction. *Bulletin of Environmental Contamination and Toxicology*, **13**: 529–549.

10. Nisbet, I.C.T. (1988) Organochlorines, reproductive impairment, and declines in Bald Eagle populations: mechanisms and dose-response relationships. In: *Raptors in the Modern World*. (Meyburg, B.-U. and Chancellor, R.D., eds.) pp. 483–490. Berlin: Weltarbeitsgruppe fur Greifvogel und Eulen e. V.

11. Helander, B., Olsson, M. and Reutergardh, L. (1982). Residue levels of organochlorine and mercury compounds in unhatched eggs and the relationships to breeding success in white-tailed sea eagles *Haliaeetus albicilla* in Sweden. *Holarctic Ecology*, **5**: 349–366.

12. Taylor, R.D. and Fynn, K.J. (1978) Fish Eagles on Lake Kariba. *Rhodesia Science News*, **12**: 2.

13. Gargett, V. (1977) A 13-year population study of the Black Eagles in the Matopos, Rhodesia, 1964–1976. *Ostrich*, **48**: 17–27.

14. Stickel, L.F., Stickel, W.H. and Christensen, R. (1966) Residues of DDT in brains and bodies of birds that die on dosage and in survivors. *Science*, **151**: 1549–1551.

15. Stickel, W.H., Stickel, L.F., Dyrland, R.A. and Hughes, D.L. (1984) DDE in birds; lethal residues and loss rates. *Archives of Environmental Contamination and Toxicology*, **13**: 1–6.

16. Anon. (1987) What's happening to Kariba's fish eagles? *Zimbabwe Wildlife*, Sept 1987: 22.

17. Douthwaite, R.J., Hustler, K., Kreuger, P. and Renzoni, A. (1992) DDT residues and mercury levels in Reed Cormorants on Lake Kariba: a risk assessment. *Ostrich*, **63**: 123–127.

18. Mhlanga, A.T., Taylor, R.D. and Phelps, R.J. (1986) HCH and DDT residues in the freshwater sardine (Kapenta) at the Ume river mouth, Kariba. *Zimbabwe Science News*, **20**: 3/4.

19. Scheuhammer, A.M. (1991) Effects of acidification on the availability of toxic metals and calcium to wild birds and mammals. *Environmental Pollution*, **71**: 329–375.

20. Brown, L.H., Urban, E.K. and Newman, K. (1982) *The Birds of Africa*. Vol. 1. London: Academic Press.

21. Cumming, D.H.M. (1981). The management of elephant and other large mammals in Zimbabwe. In: *Problems in Management of Locally Abundant Wild Animals*. (Jewell, P.A., Holt, S. and Hart, D., eds.) pp. 91–118. London: Academic Press.

22. Martin, R.B., Craig, G.C. and Booth, V.R. (eds) (1989) *Elephant Management in Zimbabwe*. Harare: Department of National Parks & Wild Life Management.

23. Greichus, Y.A. and Hannon, M.R. (1973) Distribution and biochemical effects of DDT, DDD and DDE in penned Double-crested Cormorants. *Toxicology and applied Pharmacology*, **26**: 483–494.

24. Gress, F., Risebrough, R.W., Anderson, D.W., Kiff. L.F. and Jehl, J.R. (1973) Reproductive failures of Double-crested Cormorants in southern California and Baja California. *Wilson Bulletin*, **85**: 197–208.

25. Anderson, D.W., Hickey, J.J., Risebrough, R.W., Hughes, D.F. and Christensen, R.E. (1969) Significance of chlorinated hydrocarbon residues to breeding pelicans and cormorants. *Canadian Field-Naturalist*, **83**: 91–112

26. Stickel, L.F. (1973). Pesticide residues in birds and mammals. In: *Environmental Pollution by Pesticides*, (Edwards, C.A., ed.) pp. 254–312. London: Plenum Press.

27. Hustler, C.W. (1991) *The Ecology of Fish-eating Birds on Lake Kariba with Special Emphasis on the Diving Pelecaniformes*. D. Phil. Thesis. University of Zimbabwe, Harare.

28. Skead, D.M. and Dean, W.R.J. (1977) The status of the Barberspan avifauna, 1971–1975. *Ostrich supplement*, **12**: 3–42.

29. De Kock, A.C. and Boshoff, A.F. (1987) PCBs and chlorinated hydrocarbon insecticide residues in birds and fish from the Wilderness Lakes System, South Africa. *Marine Pollution Bulletin*, **18**: 413–416.

30. Cooke, A.S. (1979) Egg shell characteristics of Gannets *Sula bassana*, Shags *Phalacrocorax aristotelis* and Great Black-backed Gulls *Larus marinus* exposed to DDE and other environmental pollutants. *Environmental Pollution*, **19**: 47–65.

31. Ratcliffe, D.A. (1970) Changes attributable to pesticides in egg breakage frequency and eggshell thickness in some British birds. *Journal of Applied Ecology*, **7**: 67–115.

32. Mendelsohn, J.M. and Sapsford, C.W. (eds) *Proceedings of the 2nd Symposium on African Predatory Birds*. Durban: Natal Bird Club.

Other Effects

9

RISK TO HUMAN HEALTH

P Goll and B McCarton

Effects in Man

The World Health Organisation considers DDT safe to people and the environment when applied in the home for mosquito control. There has been little research to support this claim.

Acute poisoning occurs when DDT is deliberately or inadvertently administered at a dose exceeding 10 ppm body weight.[1,2] The central nervous system is affected, resulting in convulsions within 30 min to 6 hours and possibly unconsciousness. However, recovery is swift and without sequelae if appropriate remedial measures are taken.

Chronic exposure to DDT after weaning appears harmless. Workers occupationally exposed to 0.25 ppm body weight/day for 25 years have shown no adverse clinical effects. Storage in adipose tissue is apparently innocuous and there in no unequivocal evidence that exposure has tumorogenic, carcinogenic or teratogenic effects. DDT can cause hepatic tumours in rats but the incidence of hepatic cancer in humans in the USA has declined steadily since 1930, despite exposure to DDT and other suspect pesticides.[1]

Babies tend to be born with slightly lower blood levels of DDT than their mothers[2] and no adverse effects have been found.[3] However, lactation is the most effective excretory mechanism of DDT and the mother sheds DDT residues at this time. DDE levels in breast milk have been linked with hyporeflexia in infants just after birth.[3] Hyporeflexia was defined if more than 4 out of 20 reflexes were low or absent as assessed by the Brazelton Neonatal Behavioral Assessment Scale. Hyporeflexia increased significantly from 3.4% at 0–0.9 ppm DDE in milk (milk fat, $n = 59$) to 14.1% at 6 ppm DDE and more ($n = 64$). The longer-term consequences of hyporeflexia are unknown but the risks should be set against those from contracting malaria due to inadequate mosquito control, or from diarrhoeal diseases due to bottle-feeding instead of breast-feeding. In 1988, for example, there were over one million clinical cases of malaria and 276 deaths reported in Zimbabwe and the true situation was undoubtedly much worse (R. Liverton and T. Freeman personal communication). Malaria and diarrhoea, and more recently AIDS, are the predominant causes of death in sub-Saharan Africa, especially in children under 5 years.[4]

REGULATORY LIMITS

Despite the apparent lack of effects in adults, regulatory limits for the intake of DDT have been set. The Acceptable Daily Intake, or ADI, is the amount of a compound which can be consumed daily, over a life-time, without risk to health. ADIs are set by the Joint World Health Organisation/Food & Agriculture Organisation Meeting on Pesticides (JMPR). The level is determined on the basis of animal experimentation and informed judgement. The maximum concentration administered daily to the test animal not producing a toxic effect is the No Observed Effect Level (NOEL). An arbitrary safety factor of 10–500 x, but generally 100 x, is applied to the NOEL, expressed in mg/kg body weight/day, to give the ADI for Man.

The Maximum Residue Limit, or MRL, is the residue concentration which determines whether a foodstuff is fit for human consumption. MRLs are set by the the JMPR Codex Committee on Pesticide Residues to allow a wide safety margin between possible intake from a 'normal' diet and the ADI.

ADIs and MRLs are reviewed periodically as new information appears and temporary levels are set where information is lacking. However, they have been criticized for their reliance on animal experimentation — and its questionable relevance to Man — and for the use of arbitrary safety factors.[5–9]

The following levels have been set for DDT and its metabolites (i.e. ΣDDT):

NOEL	0.25 ppm body weight/day
ADI	0.02 ppm body weight/day
MRL* in milk	0.05 ppm
cereal grains	0.10 ppm
eggs	0.50 ppm
fruit and vegetables	1.0 ppm
carcass meat	5.0 ppm
fish	5.0 ppm

* Temporary levels set in 1989

EXPOSURE FROM FOODSTUFFS

Fish

Residue levels were measured in commercial catches of six fish species from two sites on the lake (Table 9.1).

Samples were too small to allow statistical comparison between sample sites, although the trend suggests higher residue burdens in fish from Charara. Residue levels are broadly similar to those recorded in Berg's recent study of DDT contamination in fish from Lake Kariba,[10] although levels in Tigerfish and Kapenta at Msampa are higher

Risk to Human Health

Table 9.1 Residue levels in fish fillets and kapenta from Lake Kariba*

Species	Site	ΣDDT residue (ppm lipid)		
		n	Geometric mean	Range
Tigerfish	Msampa	2	12.4	5.2 – 29.1
	Charara	2	81.4	18.1 – 367.0
Purpleface Largemouth	Msampa	3	0.3	0.005 – 22.2
	Charara	1	15.5	
Green Happy	Msampa	1	2.5	
	Charara	1	2.6	
Mozambique/Kariba Tilapia	Msampa	3	0.4	0.005 – 8.5
	Charara	2	2.6	2.4 – 2.8
Northern Redbreast Tilapia	Charara	1	1.2	
Kapenta	Msampa	3	5.0	4.0 – 6.7
	Charara	3	3.2	1.3 – 5.5

* See Table 7.13 for species names.

than those recorded by Berg. Assuming 3.6% lipid, levels are also higher than was found in kapenta from the Ume mouth in 1984.[11]

On a worst sample basis, and on the assumption that Tigerfish muscle and whole kapenta contain 0.8%[12] and 3.6% lipid respectively, residue levels in food will be 2.9 and 0.24 ppm ΣDDT respectively, below the MRL of 5 ppm. Similarly, consumption of 200 g of Tigerfish daily would result in an intake of 0.13 mg ΣDDT, well below the ADI for a 60 kg person of 1.2 mg. Other foodstuffs are unlikely to increase the total daily intake significantly. However, the use of DDT to preserve fish from post-harvest infestation, as happens in Zambia,[13] would result in the MRL being exceeded.

Breast milk

Levels of DDE in breast milk commonly exceed 6 ppm in eastern and southern Africa where DDT is used to control mosquitos (Table 9.2) and infants are at risk from hyporeflexia. Relatively low contamination levels in Southern Province, Zambia, and one sample batch from KwaZulu, South Africa reflect the absence of mosquito control operations there. The incremental effect of tsetse fly control to the Zimbabwean and Kenyan values is unknown, but is probably relatively low unless fish are a significant part of the diet.

Table 9.2 ΣDDT residue levels (geometric mean ppm milk fat) in breast milk reported from eastern and southern Africa.

	n	Mean	Max	%DDE/ΣDDT	Source
Zimbabwe					
Mudzi District	46	30.3	81.5	41	14 and unpubl.
Binga District	15	9.5	19.6	44	15
Kenya					
Rusinga Is.	25	13.5	69.9	44	16
Zambia					
Southern Province	28	2.5	7.5	65	17
South Africa					
KwaZulu	132	15.8	59.3	54	18
KwaZulu	88	0.7	4.8	93	18

CONCLUSIONS

Mosquito control reduces mortality and morbidity in Man due to malaria. DDT remains a cheap and effective insecticide against mosquitos in Zimbabwe but residue levels in human breast milk are amongst the highest recorded anywhere. High residue levels pose no known risk to adults, but DDE levels in breast milk commonly exceed those causing hyporeflexia in infants. A study of the risks, set within the context of cost-benefit studies of mosquito control and breast feeding, should be a priority.[18] Adoption of the precautionary principle — and a switch from DDT to deltamethrin — would be a popular decision, as deltamethrin is also effective against other household pests, such as bed bugs. Operational cost comparisons indicate little difference between using pyrethroids or DDT[19] (T. Freeman personal communication 1992).

REFERENCES

1. Hayes, W.J. and Laws, E.R. (eds) (1991) *Handbook of Pesticide Toxicology*. Vol. 2. *Classes of pesticides*. London: Academic Press.

2. World Health Organization (1979) *DDT and its Derivatives*. Environmental Health Criteria 9. Geneva: World Health Organisation.

3. Rogan, W.J. Gladen, B.C., McKinney, J.D. Carreras, N., Hardy, P., Thullen, M., Tinglestad, J. and Tully, M. (1986) Neonatal effects of transplacental exposure to PCBs and DDE. *Journal of Pediatrics*, **109**: 335–341.

4. Feachem, R.G. and Jamison, D.T. (eds) (1991) *Diseases and Mortality in Sub-Saharan Africa*. IBRD, The World Bank OUP.

5. Paynter, O.E. and Schmitt, R. (1979) the ADI as a quantified expression of the acceptability of pesticide residues. In: *Pesticide residues. A contribution to their interpretation, relevance and legislation*. (Frehse, H. and Geissbuhler, H., eds) Zurich: International Union for Physical and Applied Chemistry.

6. Pieters, A.J. (1979) The setting of MRLs in food — their role and their relation to residue data. In: *Pesticide residues. A contribution to their interpretation, relevance and legislation*. (Frehse, H. and Geissbuhler, H., eds) Zurich: International Union for Physical and Applied Chemistry.

7. Tincknell, R.C. (1979) Types of pesticide residue data in foods and their characteristics. In: *Pesticide residues. A contribution to their interpretation, relevance and legislation*. (Frehse, H. and Geissbuhler, H., eds) Zurich: International Union for Physical and Applied Chemistry.

8. Frehse, H. and Geissbuhler, H. (eds) (1979) *Pesticide residues. A contribution to their interpretation, relevance and legislation*. (Frehse, H. and Geissbuhler, H., eds) Zurich: International Union for Physical and Applied Chemistry.

9. Food and Agricultural Organisation (1985) Pesticide residues in food — 1984. *FAO Plant Production and Protection Paper No. 62*.

10. Berg, H., Kiibus, M. and Kautsky, N. (1992) Pesticides in Lake Kariba. *Ambio*, **21**: 444–450.

11. Mhlanga, A.T., Taylor, R.D. and Phelps, R.J. (1986) HCH and DDT residues in the freshwater sardine (Kapenta) at the Ume river mouth, Kariba. *Zimbabwe Science News*, **20**: 46–49.

12. Matthiessen, P. (1984) *Environmental Contamination with DDT in western Zimbabwe in Relation to Tsetse Fly Control Operations*. Final report of the DDT Monitoring Project. London: Tropical Development and Research Institute, Overseas Development Administration.

13. Vaz, R. (1989) *Determination of Source of Pesticide Residue Contamination in Zambian Dairy Products*. Interim consultancy report. TCP/ZAM/8853(A). Rome: Food and Agriculture Organisation.

14. SEMG (1987) *Environmental Monitoring of Tsetse Control Operations in Zimbabwe 1986* Saarbrucken: Scientific Environmental Monitoring Group.

15. Department of Research and Specialist Services (1987) Unpublished data.

16. Kanja, L., Skare, J.U., Nafstad, I., Maitai, C.K. and Lokken, P. (1986) Organochlorine pesticides in human milk from different areas of Kenya 1983–1985. *Journal of Toxicology and Environmental Health*, **19**: 449–464.

17. SEMG (1987) Unpublished data.

18. Bouwman, H. (1991) *DDT levels in Serum, Breast Milk and Infants in Various Populations in Malaria and Non-malaria Controlled Areas of Kwazulu*. Tygerburg, South Africa: Medical Research Council.

19. Schofield, C.J. (1992) DDT and malaria vector control. *International Pest Control*, May/June, 88–89.

10

ECONOMIC ASSESSMENT

Peter Abelson*

Economists distinguish two main forms of benefit from the environment, namely user and non-user benefits.[1] The latter are often referred to as existence benefits. Damage to the environment may involve the loss of user or existence benefits.

User benefits relate to the use to which the environment is put for human consumption purposes. They include indirect effects such as impacts on human health or ecosystem effects on the productivity of natural resources. Existence values are values that humans place on the environment independently of any specific human use. For some people this includes a high valuation for the rights of non-human species to exist.

In this section, the main environmental costs due to the application of DDT in ground spraying, including the effects on food resources, human health, recreational values, ecosystem effects, and existence values, are considered.

FOOD RESOURCES

The evidence suggests that there was no impact on the fish population or on any bird or other animal resources used for human consumption.

HUMAN HEALTH

There has been much concern that DDT causes ill-health, for example carcinogenic effects or damage to the nervous system (see Chapter 9). The greatest concern related to the ingestion of DDT in food, especially breast milk, but inhalation or dermal absorption are also of concern, especially in relation to DDT use in malaria control.

---

* Overseas Development Institute, London

Unfortunately, up to now epidemiological research has been unable to provide a firm answer, one way or the other, to these concerns. However, the evidence and arguments presented point to, at most, minimal toxic effects from DDT use in ground spraying

RECREATIONAL RESOURCES

In Chapter 5 we saw that the use of DDT in ground spraying almost eradicated two songbird species and a goshawk in treated areas and that their populations were not expected to recover for 10 to 20 years. At worst, the numbers of almost half the other bird species present may have been reduced by spraying but badly affected species probably constituted less than 15% of the commoner species present in the area; none of the species at risk is endangered or endemic; and none of the treated areas is an important tourist area.

The relatively small loss in local biodiversity would doubtless reduce slightly the amenity levels of local residents. However, the impact of the DDT spraying operations monitored here on the quantity and quality of tourist experiences in north-west Zimbabwe would be negligible.

The important question arises: to what extent are the findings replicated in other areas? Douthwaite argues (Chapter 5) that serious effects in birds are related to feeding guilds and any species taking insects from tree trunks, or eating other birds is at risk. Both guilds contain a limited number of species, irrespective of habitat, and the loss of biodiversity in other habitats may be similar to that seen at Siabuwa. In any case, the loss of biodiversity due to spraying is small by comparison with changes due to natural variables, such as annual rainfall or local variations in habitat.

The Fish Eagle, a species of high recreational value, was probably the species at most risk in the aquatic environment. However, greatest breeding numbers were found in the most polluted part of Lake Kariba (Chapter 8). Usually numbers were lower where human persecution was suspected (e.g. near human settlements) or where elephants had destroyed the large trees. Again, factors other than DDT would determine tourist satisfaction.

Generally, it seems unlikely that the loss of biodiversity due to DDT use in tsetse control would significantly reduce the recreational value of most areas. However, it cannot be inferred from these observations that the use of DDT in tsetse control in areas of high wildlife value and tourist visitation would never have a high social cost. Jacquemot and Filion[2] and Diamond[3] show that birds can have very high recreational values.

Finally, mention should be made of some important indirect effects of ground spraying (and of target deployment). Many people are as concerned, or more concerned, about the intrusion of roads and other infrastructure required for ground operations in areas of nature conservation, as they are with the direct impacts of DDT. Currently, park officers in the Matusadona National Park are making strong demands that the area be cleared by aerial spraying rather than by the use of targets on the grounds that the environmental impact would be less. Access roads are visually intrusive, cause loss of wilderness character, and are alleged to facilitate illegal hunting.

Ecosystem Impacts

No significant ecosystem effects associated with the use of DDT were identified.

Existence Values

Existence values are normally associated with the threatened loss of an animal species and are greater for the more high-profile species. Such values are not applicable in the context of this study.

However, some people put a value on 'wilderness', which has had no pesticide treatment, and oppose tsetse spraying operations involving the use of DDT for this reason. Without more information it is not clear whether such Zimbabweans (and others with similar views) are opposed to tsetse operations and the change of land use that follows or to the use of DDT. These are separate issues because tsetse operations do not require DDT. My conjecture would be that if a policy decision were taken to eliminate the tsetse fly, few Zimbabweans would be willing to pay substantial amounts (in say higher taxes) to avoid DDT use solely for existence values (as distinct from user values). However, a household survey would be required to confirm or reject this conjecture.

Conclusion

The environmental damage costs associated with DDT use in ground spraying depend on the scale and place of treatment. In the current context, the damage costs were very low. However, the damage cost could be significant in areas of major wildlife concentrations, for example in national parks.

References

1. Organisation for European Co-operation and Development (1989) *Environmental Policy Benefits: Monetary Evaluation*. prepared by Pearce, D.C. and Markandya, A. Paris: Organisation of European Co-operation and Development.

2. Jacquemot, A. and Filion, F.L. (1987) The economic significance of birds in Canada. In: *The Value of Birds* (Diamond, A.W. and Filion, F.L., eds). pp. 15–22. Cambridge, UK: International Council for Bird Preservation.

3. Diamond, A.W. (1987) A global view of cultural and economic uses of birds. In: *The Value of Birds* (Diamond, A.W. and Filion, F.L., eds) pp. 9–112. Cambridge, UK: International Council for Bird Preservation.

11

THE INTERNATIONAL DIMENSION

P Goll

Global Use of DDT Today

DDT is a transboundary pollutant, borne on the wind or in water or carried by migratory animals; its continuing use is of common concern. Yet despite the interest of environmental groups, and publishers of agrochemical directories, the scale of current production and its uses globally are unknown. Enquiries directed to FAO, WHO, GIFAP, 10 regulatory authorities in tropical and sub-tropical countries, the pesticide manufacturers' associations of India and Pakistan, and 10 NGOs produced little hard information.

Only three producers are listed by GIFAP and *Farm Chemicals Handbook*[1], in Italy (EniChem Synthesis SpA), India (Hindustan Insecticides Ltd) and Indonesia (PT Montrose Pestindo Nusantara), but several 'ex-directory' producers are known. Some DDT bought in Zimbabwe was produced in France (Pechiney Ugine Kuhlmann SA) and production facilities are known in the USA (Louisiana, J. Keith personal communication), Mexico (Fertimex Corp.)[2] and Bangladesh.[3]. The 'Dirty Dozen' data sheet[4] names 11 manufacturers of DDT, but confuses manufacturers with formulators. None of the GIFAP listed producers responded to enquiries and recent production statistics are unknown, except for India, where an average of 8500 m tonnes was made annually in the three years 1986/87 to 1988/89 (Pesticides Association of India, personal communication).

The FAO collates and publishes data on pesticide consumption for agriculture in the *FAO Production Yearbook* but the statistics are incomplete. The most recent Yearbook, for 1988, lists India as the only country using DDT for agricultural purposes, consuming 3552 m tonnes active ingredient in 1987/88. However, the *Handbook on the use of Pesticides in Asia & Pacific Region*[5] names Burma, Nepal, Pakistan, Korea, Sri Lanka, Indonesia and Thailand, as well as India, as users of DDT for crop protection and Pakistan confirmed using 29–543 tonnes annually from 1985 to 1989 (Agricultural Pesticides Association, personal communication).

India has now banned the use of DDT for agricultural purposes but many countries lack pesticide legislation and regulatory controls and DDT consumption is unknown.

Cotton production accounts for over half the pesticide used for agriculture in less developed countries and DDT may remain a significant component in some formulated products. For example, Ciba-Geigy supplied 405 tonnes of ULTRACIDE, containing 35% DDT, to Tanzania for cotton pest control between 1989 and 1991.[6] Tanzania also used 287 tons of DDT for armyworm control between 1987–1989. Elsewhere, in Mali, Nigeria, Zambia and Bangladesh, DDT may be used to preserve dried fish against insect infestation[7,8] (A. Ward *in litt.*). However, samples of 'DDT' used in Bangladesh proved to be something else entirely (G.R. Ames *in litt.*)

DDT is recommended by WHO for the control of mosquitos in preventing malaria but the last review of its use was made in 1981. In that year, public health programmes in Southeast Asia, especially India, Indonesia and Bangladesh, accounted for 18000 m tonnes. In the American Region, 4700 m tonnes was used in 21 countries, especially Brazil and Mexico. In Africa DDT, totalling 1727 m tonnes, was used in 21 out of 29 countries, with Ethiopia (54%) and South Africa (22%) accounting for most. A further 1342 tonnes was used in the Eastern Mediterranean Region and 38 tonnes in European Region (Morocco); amounts used in the Western Pacific Region were negligible.

The use of DDT for mosquito control is probably in decline. Consumption in Zimbabwe averaged 225 metric tonnes per annum between 1981–89, but deltamethrin was used over half the control area in 1989/90 and 1990/91, and over all the area in 1991/92. In Brazil, mean annual consumption fell from 3060 tonnes for the period 1971–1980 to 1530 tonnes p.a. between 1980–1988. In Thailand, 1100 tonnes of DDT was imported annually in the 1970s, but by the mid 1980s funds were available to buy only 500 tonnes and the shortfall was made up with a Japanese grant to buy fenitrothion. Economic problems in Zambia ended government-financed mosquito control operations with DDT in 1985.[8]

The defence of DDT for mosquito control[9] is unlikely to stop its declining use as pyrethroid insecticides now offer politically correct, cost-effective alternatives which may be less environmentally damaging.[10]

REFERENCES

1. *Farm Chemicals Handbook* (1993) Willoughby, Ohio, USA: Meister Publishers.

2. Wright, A. (1992) *Report on a Visit to Bangladesh to investigate the Use of DDT on Dried Fish*. R1835(R). Chatham: Natural Resources Institute.

3. Ward, A. in litt.

4. Pesticide Action Network (1985) *Pesticides. The 'Dirty Dozen'*. Oxford: Oxfam.

5. Asian Development Bank (1987) *Handbook on the Use of Pesticides in Asia and Pacific Region*. Manila, Philippines: Asian Development Bank.

6. Anon. (1991) Ciba — Geigy DDT exports questioned. *Chemistry and Industry*, 3 June 1991, 367.

7. Walker, D.J. (1987) *A Review of the Use of Contact Insecticides to Control Post-harvest Infestation of Fish and Fish Products*. FAO Fisheries Circular. Rome: FAO.

8. Vaz, R. (1989) *Determination of Source of Pesticide Residue Contamination in Zambian Dairy Products*: interim consultancy report. TCP/ZAM/8853 (A). Rome: FAO.

9. Mellanby, K. (1992) *The DDT Story*. Farnham, UK: British Crop Protection Council.

10. Schofield, C. (1992) *DDT and malaria vector control. International Pest Control*. May/June, 88–89.

12

DELTAMETHRIN
A SAFER SUBSTITUTE FOR GROUND-SPRAYING?

R J Douthwaite

Ground-spraying trials with the synthetic pyrethroid insecticide deltamethrin were conducted in the Omay Communal Area in 1989 and 1990. The first operation was monitored by a local consultant[1] and the second by a team from the Natural Resources Institute.[2]

Terrestrial Effects

Residual deposits of deltamethrin (0.05% s.c.; Glossinex 200) were applied to tsetse resting sites at an overall rate of 2.5 g/ha. Spray droplets were collected on water-sensitive papers to assess their dispersal. Their fallout was strongly influenced by wind, but adjustment of the spray nozzle and operator targetting were also important. Contamination was greatest within 3 m of the point of application but some droplets were carried up to 10 m. Treatment of the vegetation beside a stream led to contamination of the water. Residues averaged c. 20 mg/m^2 on tree bark, much less than the 90 mg/m^2 recommended for effective control of *Glossina pallidipes*. Eighty to ninety per cent of the deposit was dissipated within two months.

A wide range of invertebrates were affected on mopane tree trunks. Nine weeks after spraying, numbers of planthoppers (Homoptera: Fulgoroidea) and silverfish (Thysanura: Lepismatidae) populations were still significantly reduced (Figure 12.1). There were also indications that spiders, particularly Salticidae and Lycosidae, and an arboreal grasshopper (Orthoptera: Thericleidae) were affected over a two-month period.

No effect was detected on the ability of the ground-dwelling ant fauna of mopane woodland as a whole to locate favoured food baits. However, the ponerine ant *Platythyrea cribrinodis* did find fewer baits in the sprayed area immediately after spraying, whilst foraging efficiency in neighbouring unsprayed habitat was unaffected. This effect was only apparent in the short-term.

DDT in the Tropics

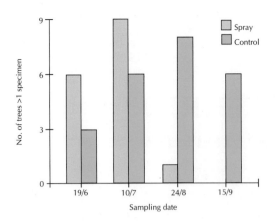

Figure 12.1 Numbers of Fulgoroidae in sprayed versus unsprayed areas

Large numbers of the arboreal ants, *Camponotus* spp. and *Crematogaster* sp., were 'knocked down' by spraying but their fate, and the fate of their colonies, were not studied.

Studies on lizards, birds and mammals were compromised by prior treatment of the study areas with DDT. However, there was no evidence of an immediate effect on the numbers of skink *Mabuya striata*, White-headed Black Chat or night ape *Galago senegalensis*.

In the longer term, numbers of White-headed Black Chats were probably less affected by the treatment with deltamethrin than by a second treatment with DDT. Between September 1989 and June 1990, numbers fell by 55% in the deltamethrin treatment compared with 75% in the area treated with DDT, but only small samples of birds were involved. Decline in the former area was probably due to treatment of the area with DDT in 1988 although a contributory, indirect effect, due to deltamethrin reducing food availability, cannot be ruled out. Further study of feeding success and survival of insectivorous birds in a treated area are necessary to confirm the greater safety of deltamethrin over DDT.

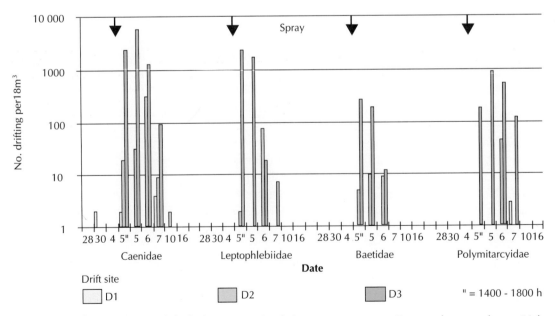

Figure 12.2 Changes in aquatic drift of Ephemeroptera in relation to spray treatments (Two samples were taken on 5 July, immediately after treatment)

Deltamethrin: a safer substitute?

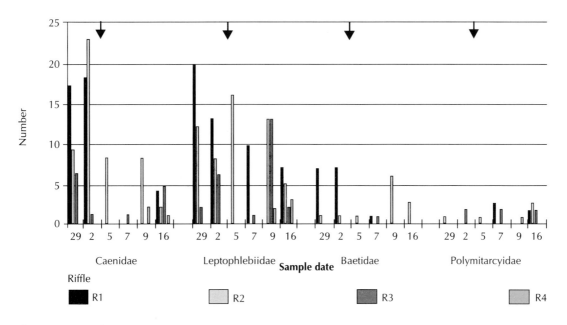

Figure 12.3 Numbers of benthic Ephemeroptera collected from 12 similar-sized stones per sample site

Aquatic effects

Spraying increased the natural downstream drift of aquatic invertebrates by up to three times for several days. Mayflies (Ephemeroptera) in particular were affected but residual populations were not seriously impoverished (Figures 12.2 and 12.3). Behavioural changes and some mortality amongst surface dwellers (Gerridae, Hydrometridae, Gyrinidae and Notonectidae) also occurred but no mortality of fish or crabs was seen. Most streams in Zimbabwe dry up each year and the effects of deltamethrin on aquatic fauna should be viewed in this context.

References

1. Hustler, K. (1990) Report on effect of insecticide spraying for tsetse fly control in the Mola and Tiger Bay areas of Zimbabwe in 1989. *Report of the 3rd Annual Review Mission, Environmental Impact Assessment of Ground-Spraying Operations against Tsetse Flies in Zimbabwe*. Chatham: Natural Resources Institute.

2. Lambert, M.R.K., Grant, I.F., Smith, C.L., Tingle, C.C.D. and Douthwaite, R.J. (1991) Effects of deltamethrin ground-spraying on non-target wildlife. *Technical Report No. 1, Environmental Impact Assessment of Ground-Spraying Operations against Tsetse Flies in Zimbabwe*. Chatham: Natural Resources Institute.

Roma

Fuksas's atelier in Rome is situated in a Renaissance building, whose present configuration dates back to the second half of the XIX century, in *Piazza del Monte di Pietà*, near Campo dei Fiori.
The atelier has four floors and hosts offices, product/interior design rooms, model making room and meeting rooms. A lift with an iron platform, crystal walls and natural steel fixtures connects the first floor with the "aquarium room", where meetings take place. In some parts of the atelier, the lime putty treated walls have been renovated in order to bring out the plaster and paint layered by time. Hanging on the walls are several models shielded in showcases as well as paintings and sketches by Massimiliano Fuksas. Among the furniture of the atelier, the table "Tommaso" for Zeus, the chair "Ma-Zìk" for Saporiti, the office chair "Bea" for Luxy, the corporate table "Mumbai" for Haworth Castelli, the chair "Bianca" for Zeus, the table "Biennale Collection" for Saporiti e the "Panton Chair" di Verner Panton. In Fuksas's personal room, a wide iron sliding door acts both as background and support for sketches, images and pictures of projects. The exposed ceiling is made of white painted wooden truss. Among the furniture: lounge chair by Eames, the chair Ant by Jacobsen, table series B637 by Hein, Mathsson, Jacobsen. Near Fuksas's atelier, in *Via di Santa Maria in Monticelli*, there is the exhibition and deposit area which houses, among the other things, the models for important projects. A stairway links the 3 levels of the building, which enshrines in the basement the ruins of Roman baths.

FUKSAS object

This book is a comprehensive look of the design by Doriana and Massimiliano Fuksas, a multi-scale and stimulating proposal, produced from their continuous curiosity on the contemporary medium.

Our first step is outlining their production in <u>scenography and exhibitions.</u>

LESS AESTHETICS, MORE ETHICS

VII International Architecture Exhibition in Venice, Italy, 2000

The urban transformation occurred in the last thirty years has no equal both in terms of dimension of the phenomenon and the size of the areas involved. The first idea was to use the 2000 Biennial as a workshop for analyzing and trying to give an intelligible shape to the new planetary dimension of urban behaviours and transformations. The considerations, investigations and intuitions over the evolution of the cities took shape as a need for "something else". "Something else" from architecture - whose troubled life we try to prolong every day – which we share our entire life with, and "something else" from successful architecture schemes: it was about recovering the awareness according to which the quality of architects and works was no longer enough. At such point, the 2000 Bieannial edition had a main theme.

"CITIES: LESS AESTHETICS, MORE ETHICS"
The theme of the seventh Biennial of Architecture, tries to communicate the deep unease of fast-transforming societies, where the data and reference points of an architect have utterly changed. The exhibition, set at *Le Corderie*, draws the visitor's attention to the "big" 280 X 5 m screen whose images posit questions about megalopolis, areas contaminated by contradictions, conflicts, pollution, refugees' dramatic condition, about new social aggregation centers like stations, airports and shopping centers, as well as a series of interviews with fifty architects. For the first time, the Venice Biennial of Architecture uses simultaneously *l'Arsenale*, that is *le Corderie*, *le Artiglierie* and *le Gaggiandre*, in addition to *i Giardini*, the exhibition's traditional location.

A homeland to every seaman and castaway. Getting to think to a real not virtual place to be dedicated to Peace is a feat of great responsibility and deep ethic intensity. Peace is a soul place, is an aspiration, is tension and utopia. Indeed, it represents that desire for completeness and serenity that can be conveyed by space and architecture. We thought of a stratification, a construction representing time and patience. The Center for Peace made for the 2000 Biennial is a parallelepiped obtained from stratification of irregular concrete and glass surfaces resting on a small monolithic platform. It is about a stratification made by means of alternating "matter", representative of the places where suffering is widely known. The inside becomes a great square for welcoming visitors, where size and height, together with the light filtering from above, help to forget the anxiety of the world and positively encourage people to meet and talk to each other.

INTERNI D'AUTORE IN PIAZZA
LA CAPSULA ABITABILE   TRIENNALE, MILAN, ITALY, 2002

The installation is characterized by the notions of lightness and transparency, obtained thanks to a compositive sequence of quasi-cinematic montage, made up of overlapping layers, distinct matrix and communicative screens, which as a whole offer a message that changes according to the observer's viewpoints. Walls and their transparent cover, composed as a sandwich, contain sachets of freeze-dried food, arranged as bricks and filtered by the writings "Water" and "Food": primary elements of subsistence reduced to a publicity sign. The sachets of food, which as an industrial product is translated into an impersonal and infinitely serialized element, wrap visitors and inhabitants - two astronauts who observe impassively their present - in a distressing domestic capsule, predicting a near future. The immateriality of the construction is also underlined by the mirror floor as well as the dangling, within the room, of further processed foods, now in coloured wrappings, alongside screens showing images of the planet earth in its natural state, before the anthropization and the domestication-destruction of the landscape and territory took place. The perception of nature and the animal world at the wild state contrasts with those summarized in the project: foods, and thus nature, no longer recognizable in their original shape, but turned into sachets, that looks identical, and domestic animals that are but robots.

FORMA
LA CITTA' MODERNA E IL SUO PASSATO
COLOSSEUM ROME, ITALY, 2004

A path inside the Coliseum's II level to try to chronologically revive, though briefly, the Roman History from the first century to 2004. Some of the archeological works on display belong to the Coliseum's permanent collection, while others had never been exhibited before.
A large 3-metre-high wavy ribbon – raising 1 metre from the ground and divided into three sections – wraps around the Coliseum's fornixes. Historic materials like engravings, paintings, sculptures belonging to other Italian museums are projected on the first section of the ribbon-display. On the second one, other videos show the transformation of The Fori's central area from 1500 until our planning proposal for the renovation of the entire *Via dei Fori Imperiali*: from *Piazza Venezia* to the Coliseum. The lighting of the exhibition uses two types and colors of light: a blue one on the floor and vaults (which by no means interferes with the displayed marbles) and a white one to accurately light up sculptures and archeological relics. Interviews to personalities from the culture, cinema, sport and music world are shown on ten different plasma screens. Two more wide screens show brief clips dedicated to the movies and actors that have interpreted ROMA....
Getting to the last section: the reconstruction project of the whole central axis area going from *Piazza Venezia* to the Coliseum.
The same blue light of the exhibition set-up expands along the central axis of the *Fori*, underlining and emphasizing the wound represented by *Via dei Fori Imperiali*. At night, a blue light of variable intensity becomes a pulsation, a heartbeat coming out the grills set on the arches of the II level and spreading all along the Coliseum's view: it is the sign that the city keeps on living and the only way to keep past architectures and monuments alive is to use and live them in a contemporary way.

Roma, Colosseo, Mostra "Forma. La città moderna e il suo passato.". Progetto a cura di Adriano La Regina, Massimiliano Fuksas ① e Doriana Mandrelli Fuksas ②. Committente: Soprintendenza Archeologica di Roma ④. Consulenza tecnica per l'illuminazione: iGuzzini ③. Prodotti: Proiettori Metro, design Bruno Gecchelin.

MFuksasD
unsessantesimodisecondo
MAXXI Museum, Rome, Italy, 2006

One sixtieth. A fraction of a second. The idea penetrates within such a little lapse of time. The idea-strenght. Seeing a film backwards, from the end to the beginning; this is the idea of the exhibition at the MAXXI Museum. Starting from the completed work and showing all the preceding phases in reverse order: the first ideas consisting of rapid notes scribbled on any kind of support, the many studio models produced at almost the same speed of thinking, the experimentation of different techniques and diverse materials, from the continual modifications to the point of synthesis, up to when the idea materializes. This is what we want to show, the passage of an idea that materializes in a sixtieth of a second. The exhibition retraces altogether the works carried out by Studio Fuksas, starting from the latest designs and going backwards up to the early projects. Along the exhibition itinerary, visitors can grasp the creative process pervading every single work, from conception to realization. Models, sketches, paintings, photographs and videos show the various stages through which the plans develop.

Fuori Salone
Internazionale
del Mobile,
Interni-Decode
Elements
**<u>Showcase for
La Rinascente,</u>**

Piazza Duomo_Milan_Italy_2007

The cabinet, entirely upholstered with reflective materials, stages a sculpture composed of fourteen curvilinear sections of varnished metal sheet, arranged into the space to form the volume of a cloud. Golden and gold-plated silver jewels from the "Islands" collection, created in cooperation with Mimmo Paladino for Short Stories, float around the red whirling sculpture as shining cosmic presences

Fuori Salone Internazionale del Mobile
Interni — Decode Elements

Beside the eclipse

Castello Sforzesco_Milan_Italy_2007

The installation is composed of an 8-metre-wide oscillating circular platform upon which rotates a glass-covered platform. Beneath the walkway, coloured liquids encased in an interactive surface moves with the weight of the visitors, thus creating dynamic shapes. A walk on earth, the azure planet. A tilting platform to remind us of the uncertainty of our condition, a dream of transparency to finally see the heart of things. Going to the center for a moment without passing through the surface. Our responsibility in the shape of an abyss. The imprisoned earth is suffocating, we are heavy, we make the difference, we deform its crust, we exploit and rebuild it. A window on the glassy smooth curve, a trapdoor, a secret. Tautologies apart, to see makes the difference as a premise for any form of knowledge. The basements of our soul, a planet that is evaporating.

Multi-Hábitats: La vivienda polivalente
Prototype of house_2 x 1
International Construction Fair__Barcelona_Spain

The aim of this scheme is to foster new ways of thinking about the contemporary habitat. Envisaged as a floral based system, the project suggests a home-living/housing device resting on two petal-units. The first rigid one houses the circulation area while the second one, more flexible and, with variable dimension, defines the living space. The latter is articulated through a large dorsal element functioning at the same time as furnishing element and mobile partition. The 90-degree rotation of the central "spine" allows different home-settings through an either longitudinal or transversal breaking up, depending on the case. The planning proposal aims to create a scenery "that transforms itself" depending on the moment, the time of the year or the personal interests in daily life. One space or two, at the same time. Two for one. On an urban level, the scheme sets an height challenge for itself. The vertical combination of the 70 and 40 square metres overlapping and out-of-axis units results in a bunch structure around tower-shaped elements that function as central core and can be arranged irregularly in terms of both density and height whilst also being able to redefine some central or peripheral areas of the city.

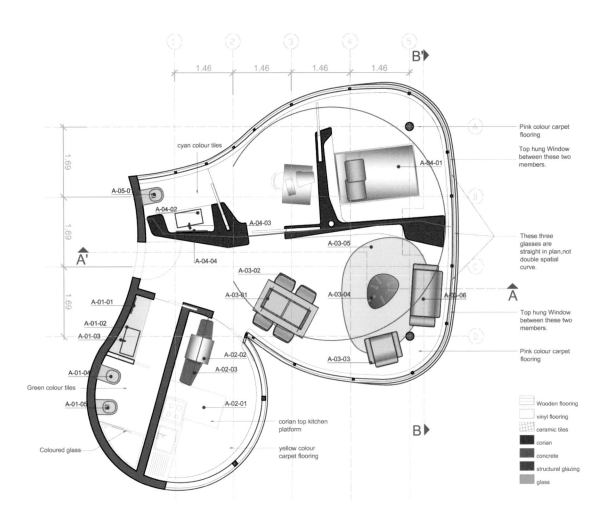

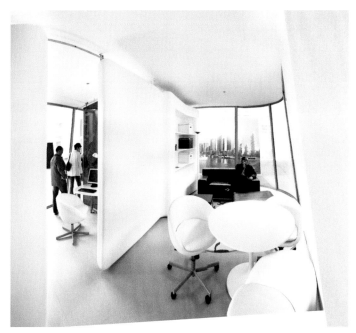

A HOME FOR ALL
DEEP PURPLE
TRIENNALE_MILAN_ITALY_2008

Three floors, bright, furnished, with garden. No garage, just an inflatable punt. This is our house for everybody. This is a positive reflection about a too often negative reality. The earth is getting hotter, more discontented, polluted and restless. Unpredictable nature. Sudden waves, drought and earthquakes.

Skies are about to rain for days and days, stars are about to fall too much or too little, while satellites – forgotten in their spatial orbits - suddenly perforate the atmosphere, drawn to the ground by mysterious forces. In this status of constant emergency, where every day seems to be (and maybe is) worse than the previous one, it is important to decide which decision to make. Giving in to the drift or turning it into a push, an event (though inescapable) that maybe will eventually lead us to discover a new earth. A new country. A place that nobody has already named. Where things haven't got a name, because there's no need for it. But the journey could be too long, even solitary. Impervious perhaps, formidable. Extreme.

We will only have the time to take the essential, maybe not even that. At this point our house, weightless as an orange lantern light that floats into the ocean, will be the only thing that we have. One coral surface, a sign. A white-hot fire that removes insecurity and burns doubts. It protects the wonderfully human desire to carry on, to go always further. Our house is a strong and intelligent structure, that has drawn only the best from science and technology. Osb (oriented strand board) to divide internal space, atex5000 silicon-impregnated fibreglass to protect us from the outside, inflatable furniture where to rest and dream of a landing place. The arrival. The Earth how we have never seen it. Everything is arranged within 5 x 5 metre spaces, equipped to ideally face every situation without losing the pleasure for things. The pleasure that comes from things. From their beauty, not only aesthetic, but from the process that has generated them. From their reason and necessity.

MADRE Museum
Naples_Italy_2008

Napolincroce

An installation, set around the altar of the Church Donnaregina Vecchia within the Museo Madre, composed of a "forest" of chestnut trunks kept together by nails and pickets.

The light plastic boxes are boldly in scale to the Corderie of the Arsenale. The film that can be seen inside them does not take on any voyeuristic undertone but tells the story of young people's everyday life and their penchant for consuming and accumulating huge quantities of waste.

Home, container, refuge, a space where life takes place. A man and a woman meet each other: They look at each other, they speak, they eat, they see friends…they love each other. They eat a lot, they drink, they overstock food and tableware, making a lot of waste. Behind their shoulders a big TV screen keeps on showing news. A dog is rummaging through the rubbish. Into the 3 green, translucent, light and sexy boxes stories of daily life repeating time and again, a baby assures continuity. They are experimental houses, easy to be fixed, set up into the Corderie in Venice. If necessary, they can be moved to any other space, whenever an emergency may occur. That is what "Out there: Architecture Beyond Building" means to us: beyond building, the human being… life and its contradictions.

Kensington Gardens

"Out There: Architecture Beyond Building"
XI International Architecture Exhibition in
Venice_Italy_2008

Scenography for Medea and Edipo in Colono

Greek Theatre_Siracusa_Italy_2009

The horizon has been the inspiring element of reflection for the design of the set. A very simple horizon that speaks of catharsis, that is the landscape, place and action for the conscience of the spectators, not today but yesterday.
As the landscape and scenery have changed over time, the intention contained in the scenography is the reconstruction of a lost horizon. Both for "Medea" and "Edipo" performances, the scenography is constituted by a concave mirror metal blade (w 22,39 x 10,91 x 14,80 mt) that reflects what is happening around, which involves the public, reflecting and inviting reflection. On the floor of "Medea" scenography there are letters from the Greek alphabet and symbols. In the "Edipo" scenography there is a hill made of salt.

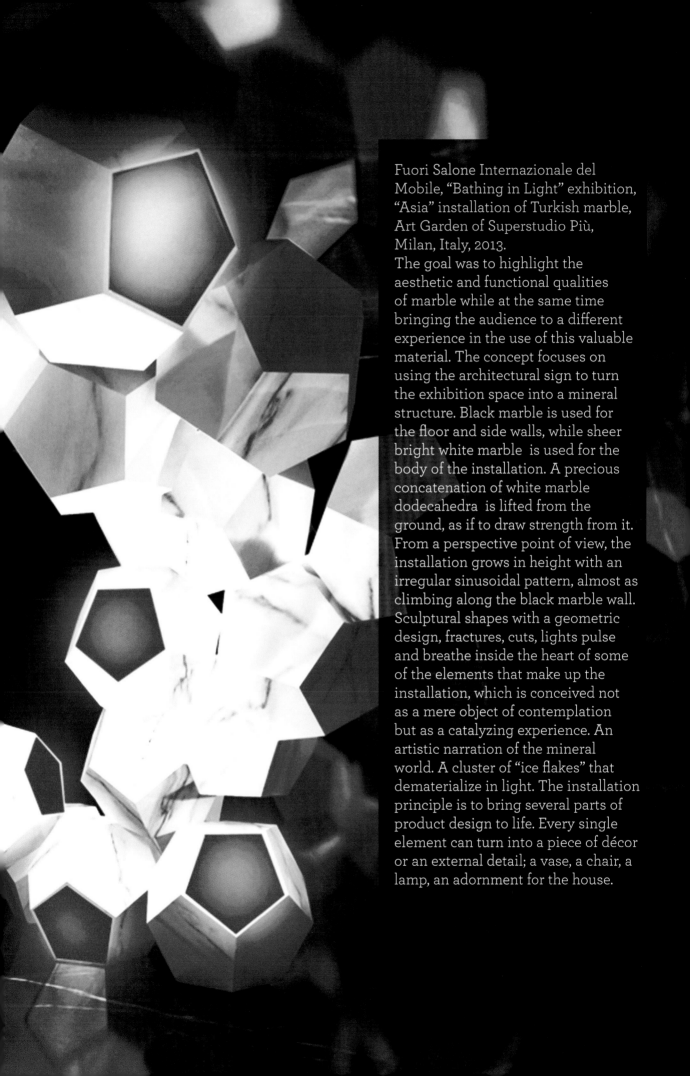

Fuori Salone Internazionale del Mobile, "Bathing in Light" exhibition, "Asia" installation of Turkish marble, Art Garden of Superstudio Più, Milan, Italy, 2013.

The goal was to highlight the aesthetic and functional qualities of marble while at the same time bringing the audience to a different experience in the use of this valuable material. The concept focuses on using the architectural sign to turn the exhibition space into a mineral structure. Black marble is used for the floor and side walls, while sheer bright white marble is used for the body of the installation. A precious concatenation of white marble dodecahedra is lifted from the ground, as if to draw strength from it. From a perspective point of view, the installation grows in height with an irregular sinusoidal pattern, almost as climbing along the black marble wall. Sculptural shapes with a geometric design, fractures, cuts, lights pulse and breathe inside the heart of some of the elements that make up the installation, which is conceived not as a mere object of contemplation but as a catalyzing experience. An artistic narration of the mineral world. A cluster of "ice flakes" that dematerialize in light. The installation principle is to bring several parts of product design to life. Every single element can turn into a piece of décor or an external detail; a vase, a chair, a lamp, an adornment for the house.

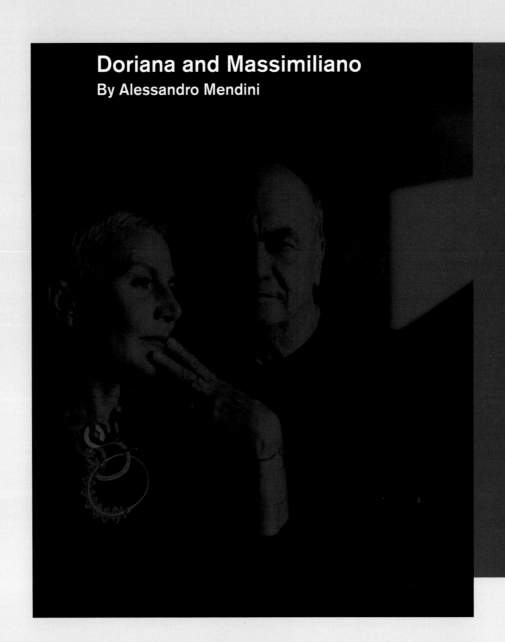

Doriana and Massimiliano
By Alessandro Mendini

0

I would like to start by observing how the design made in the Fuksas office takes its character from combining and contrasting two distinct types of poetic expression: abstract and narrative in Doriana Mandrelli's approach, and intuitive and textural in Massimiliano's. Brought into contact, these two all-encompassing means of seduction produce a world of objects that are autonomous and original, when compared to the standard procedure ingrained in the logic of Italian design.

Mandrelli and Fuksas's design originates in the collision and encounter between two forces that have entered the design profession from experiences far removed from design, and it is precisely this procedural unconventionality that lends the objects their character.

The background that the couple uses as a horizon for their objects features traces of their respective cultures and peregrinations. It contains Massimiliano's love of Jean Prouvé's mechanicality, Doriana's love of the archaicism contained in chiseled gold enveloping the body, and the science fiction-like mirage that both of them foster for an ethical and ceremonial liberation of humanity.

1

As a result, the orthodox history of Italian design has difficulty containing the Fuksases' products within the limits of its canonical rules.

Italian "Bel Design", Milanese by birth, is analytical, pernickety, precise and slightly asphyxiating, besides being industrial and pragmatic. (Only the idealism of "Contro Design" was able to redeem it from this stylistic realism by injecting it with heightened emotive exuberance). The design of the Fuksases does not fit into this mould. In what is perhaps a move of snobbery, their design is not even interested in knowing what the typical methods of Italian design are.

From the beginning, it shrugged off the dose of Calvinism contained in a design history that is more graphic than spatial. The cosmopolitan attitude of Doriana and Massimiliano; their prevalent humus being architectural in origin; their artistic sensibility favoring modern Roman painting and contemporary French art; and their ever-different and well-structured modus operandi are all elements determining the brilliant uniqueness of the "repertory" of their objects, installations and settings.
Families of household tools with a high degree of elegance and spatiality go hand in hand with the modeling of technical and modular seating in the style

of Roger Tallon; huge hollows of commercial space endowed with slick and striking dynamics; immense nocturnal settings more similar to land art than theater; liquefied sinks; matte gold jewelry with swollen organic shapes reminiscent of cuttlefish bones. Then tables, handles, lamps, lampposts and urban appliances.

The performance-factor of their design is obvious, where each object/personage joins a precise hypothesis that is warm-blooded, type-specific and experimental.

This meta-language includes American-style inventions that seem to pay homage to Raymond Loewy's pop or Eero Saarinen's naturalism. Clouds, grids, membranes, and bubbles large and small are rolled up, curved out and under, and placed in the world's geography at all latitudes, with great self-assurance.

In a pendulum swinging between sophistication and brutalism, primary colors and gloss, new materials and polychrome light-effects, robotic suggestions abound. Their image is perfect to furnish the undifferentiated, infinite surfaces of enormous buildings in a galactic cosmos.
And this brings us to the next observation: that of the clash, or rather the encounter residing in the

office: that of architecture (macro-organisms) and that of design (micro-organisms). To look at the objects designed by this office is sometimes like looking at Massimiliano's architecture, only that it's been made small and seems far away because you're holding the binoculars backwards. The large becomes minuscule, the monumental becomes miniature, the hostile becomes friendly by means of radical genetic transformation.

The status of the architecture changes. Instead of a protective space for design, it becomes design itself: domestic prosthetics.

Except for a few miraculous cases (such as Josef Hoffmann, Peter Behrens, Le Corbusier and Gio Ponti) the conflict between architecture and design inside the same office has always remained unresolved. It did not get transformed into synthesis, but into complexes that remained hanging. One of the two approaches inexorably prevails over the other. But in the positive, well-resolved case of Fuksas and Mandrelli, that miraculous phenomenon occurs.

The two competing mentalities are integrated by a magic wand, by a single unitary desire: the possession of one unique design utopia, elaborated by two different visions. Several great architects, Marcello Piacentini for example, kept their design within the confines of basic fixtures such as built-in furniture, handles and lamps that were strictly meant to further their architectural aim.

Others, such as Jean Nouvel, amplified this to include a few pieces of office furniture, but still in direct service of the spaces of their architecture, as if they were an ergonomic consequence, the natural development of a sub-module.

As for the Fuksas-Mandrelli team, the design issue has become autonomous. It is outlined with its own detached character, because it takes on a different territory from that of space: the territory of anthropology. Here, the parameters of judgment slide toward humanistic breadth and psychic involvement of an entirely different nature: that of the consumer, the person faced with innumerable objects and tools to use as scenic instruments.
These aesthetic presences are placed in front of us, speaking to the existential responsibility and irresistible narrative power that both they and we take pleasure in conveying.

5.

TEA COFFEE TOWERS & MORE

WORKS FOR ALESSI 2003-2012

"Architectura magistra omnium artium"

In the realm of contemporary architecture and design the case of Doriana and Massimiliano is, in my view, a unique example of a cultural and personal osmosis that while usually tends to rub off the identity of one of the partners, in their case it reinforces and improves the overall outcomes of their projects.

It is a given that the tradition of Italian design has always had a strong connection with architecture. From the Fifties onwards there has never been, with few exceptions, an important designer who had not formerly been an architect. Once it was believed that in Italy there were not enough commissions and for this reason architects were somehow forced to take up design in order to bring home some work (I remember in this respect the regret of Ettore Sottsass, or of Aldo Rossi who used to complain: "instead of inviting me too often to chair architecture juries, they should give me some work"). Yet, I rather believe that the reason lies in the education given by the universities of architecture compared to the schools of design: an architect has a wider, deeper, humanistic view of the world and of his job, whereas a designer still often falls victim of the Ulmian vision of form follows function.

And Alessi too is no exception: although not ostracising the designer's canonic discipline, it is a given that at

Alessi's a majority of the most interesting works have been realized by architects: from Caccia Dominioni to Zanuso, to the Castiglioni brothers, Sottsass, Mendini, Dalisi, Rossi, Branzi, Lissoni, let alone the foreign architects: Venturi, Graves, Botta, Foster, Ito, Chipperfield, Kaplicky, Alsop, Sanaa, Hadid, Arets...

In this context Doriana and Massimiliano represent a most significant experience. Their partnership denies the false belief that architecture is masculine while design is feminine: belief which is stupid as well as rebuffed by half a century of statistical data, yet Achille Castiglioni used to reproach me jokingly because I asked him to design feminine home objects that he had never used nor could he use.

Doriana and Massimiliano are two talented creators who seem to have separated their fields of action: obviously they keep exchanging ideas and move from a shared poetical matrix, but one is mainly devoted to big projects, the other to small ones. Since my profession concerns small projects, I work nearly exclusively with Doriana. She has a beautiful personality: she is strong, insistent and inflexible when necessary, but she is also tender and sensitive as none of the male designers I have worked with. She has enriched my catalogue with some of the most beautiful outcomes of recent years.

Alberto Alessi

A continuous, flexed, fluid plane. Material is the only protagonist. A "sheet" of delicate porcelain, of silver, plastic or glass enclosed on itself asymmetrically it is set on a sinuous plane. Together with the two coffee cups that bring their same name, E-LI-LI, the two flowerpots are the first series of products that derive from the intense study of the previous design project called "Tea & Coffee Towers" in 2003. Both the vases can be used together or separately. The ceramic and the steel, materials that is difficult to model, are used in these two vases as if they were a paper sheet or a skin fragment.

E_LI_LI 2005

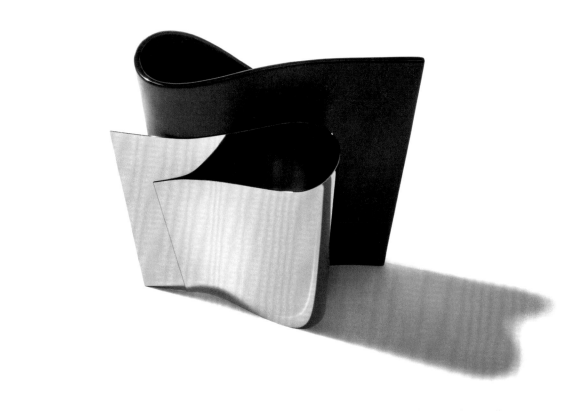

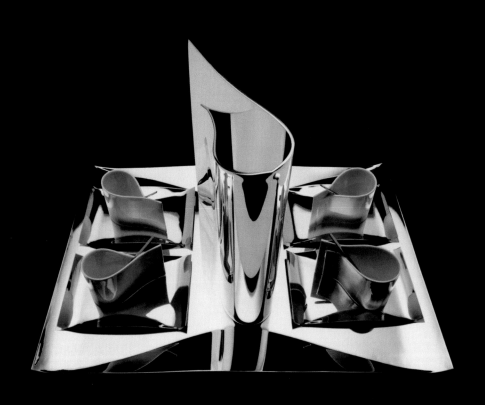
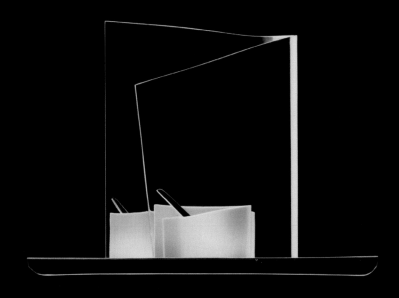

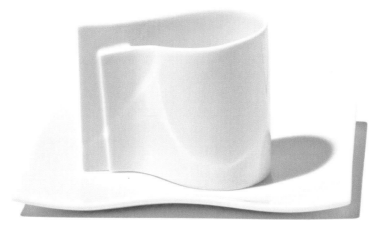

Coffee Cups

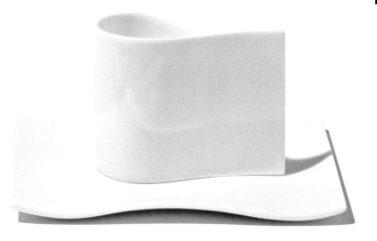

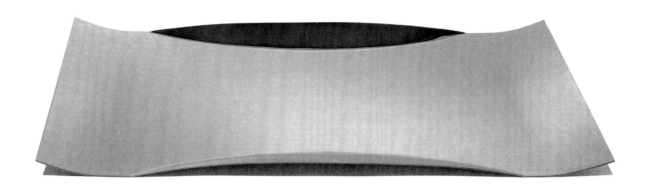

mao mao 2005

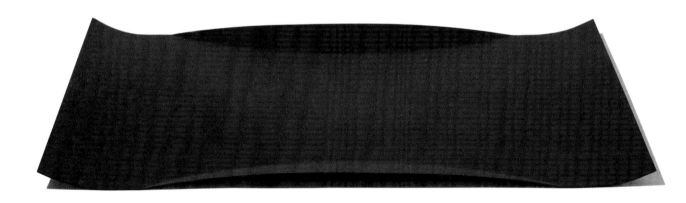

The tray in stainless steel represents the move to mass production of the concept created in 2003 for the "Tea&Coffee Towers" project.

Colombina Collection 2007-2010

The "Colombina collection" presents itself as a project distinguished by a sculptural approach: new, pleasant, surprising plates, containers and elegant cups, made from different materials and in different colours. Yet, behind this project there is a profound knowledge of table rituals: as a matter of fact the proposal is very close to the revival of the so-called "French" dinner service (ancestor of the modern buffet). It is not a mere gratuitous historicist recycling, but on the contrary is a complete set open to the evolution of the modern concept of dining, thanks to its jovial component and its aptitude to transform and modify the composition of elements on the table. One piece contains the other and all can be freely combined, like an inviting petals' stack to be peeled one by one according to the guests' preferences and, obviously, the dinner options.

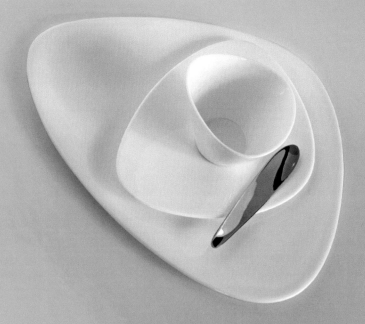

Colombina Collection

Colombina Collection

Black, white or red melamine tealight holder.
Pretty, flexible and fun, completing the "Colombina collection" family.

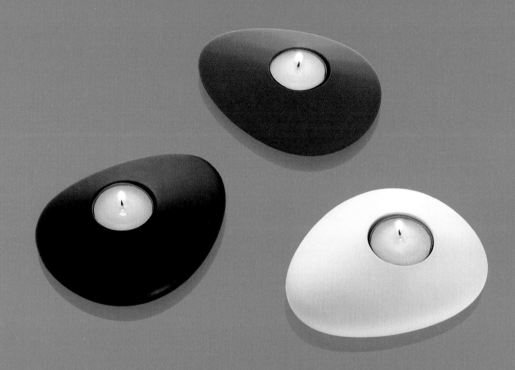

Salt & Pepper

Colombina Collection

Colombina Wall mirror with black, white or red glass frame and placemat in melamine, red or black, and tray in stainless steel.

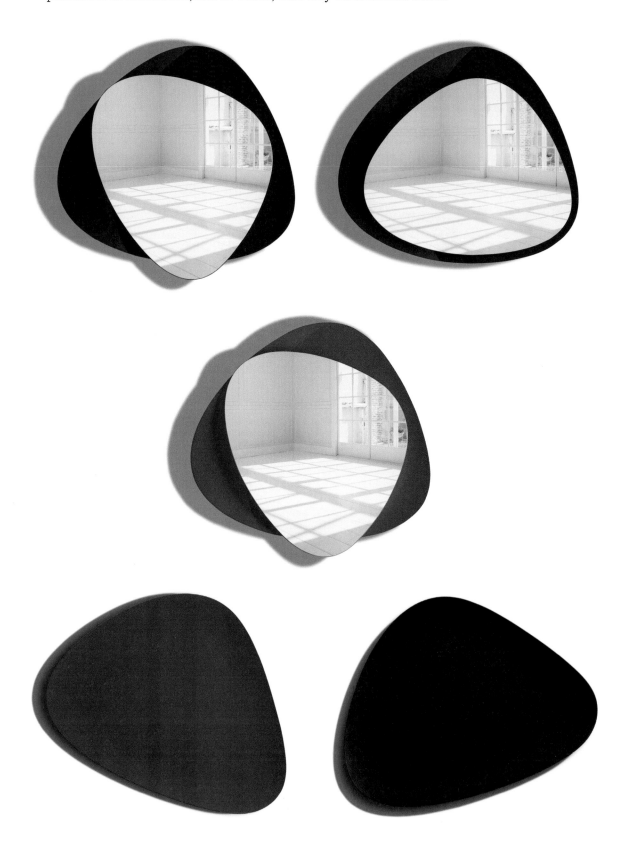

Colombina Collection

Crystal-wine, water, champagne, liqueur glasses, part of the "Colombina Collection" complete tableware set, which embodie the movement of contemporary dining towars a more flexible, dynamic and playful approach. All with different finishes, the pieces can be freely combined with each other.

Sugar

Colombina Collection

A complete collection of table cutlery made of 18/10 Stainless Steel is presented in 2007. We mention the three-prong forks with their softly curved lines and the mocha coffee spoon, the ice cream spoon and of the hors-d'oeuvre fork with their particular shapes.

Table Cutlery

TABLE SPOON FUKSAS
12/05/06
TOLL. 2,2 MM.

DESSERT SPOON FUKSAS
12/05/06
TOLL. 2,2 MM.

TABLE KNIFE FUKSAS
12/05/06
TOLL. 2,2 MM.

TABLE FORK FUKSAS
12/05/06
TOLL. 2,2 MM.

COFFEE SPOON FUKSAS
12/05/06
TOLL. 2,2MM

Table Cutlery

Zouhria

Large ceramic vase. "Zouhria," which in Maghreb Arabic refers to a "flower vase", it's a limited edition home accessory made in 99 numbered pieces and 9 artist's trial pieces. It has a smooth rounded shape recalling the desert stones sculptured by the Maghreb wind. Designed in 2010

baby

Brown or white porcelain citrus bowl. Light blue inside.
The pierced surface of Citrus basket Baby creates a sort of grid through which the fruit inside is visible.

Brown or white porcelain tealight holder. Red inside. When placed inside Baby, the T-light's luminosity coming through its perforated surface creates a striking effect of lights and shadows on the living room or kitchen walls or in any other space surrounding it. Designed in 2011.

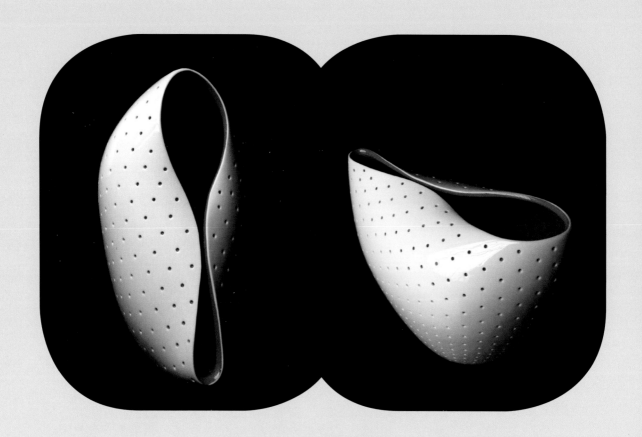

ALDO

Black or white porcelain flower vase. Orange inside. The "ALDO" Vase, a further piece added to the object series for the living room, is made of porcelain. The external surface is riddled to let the different colour of the inside visible. The top part of the vase is flattened to form two "openings" able to host floral compositions or single stems. Designed in 2012.

In real life it is unusual and rather difficult to see pieces falling in place by themselves and dovetailing neatly.

It is usually a pretty strenuous activity to make them fit or work with each other. It rarely happens that a sort of destiny's mikado allows people to find and recognize themselves.

Yet, sometimes it happens, because of an inexplicable combination of factors or merely by chance.

And when it happens, when such a precious event arises, the most interesting and rich encounters occur.

Those that don't need any excess or extra words to function as a perfect devise.

Essential, precise, yet poetical.

Encounters allowing to produce ideas and create new solutions together.

It happened to me with Doriana, I hope she feels the same.

Giorgio Armani

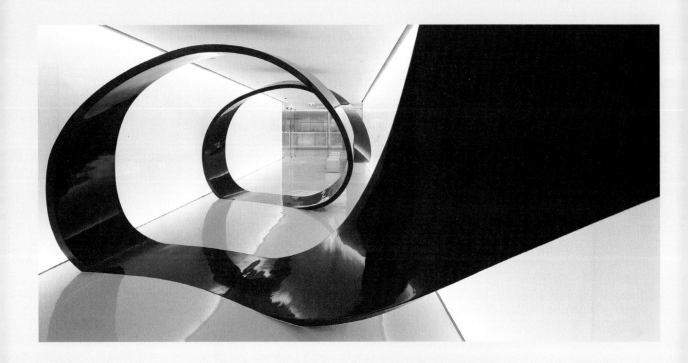

ARMANI Chater House
Hong Kong _China_2002

The project for the Emporio Armani in Hong Kong focuses more on voids than on solids. Fluxes, rather than decoration, are the real inspiring motifs. The concept is a development of fluidity, reached by studying the casual motion of visitor's trails as well as by the construction of the exhibition space on the base of such invisible routes.
The exhibition area for fashion develops within luminous paths running between two carved, well-lighted glass walls that house clothing and accessories and mirror themselves on a bright resin floor which doubles their dimension and multiplies their reflections.
From the Emporio you can have direct access to the restaurant, characterized by a red varnished stripe that starts from the floor, lifts up to form a table, descends again to accommodate the restaurant space, then closes and intertwines to become the counter, and eventually forms a tunnel/spiral which draws the space up to the hall.
The fiberglass stripe is 105 metres long, 8 metres high and 70 centimetres wide (they are two stripes joining and merging together in the centre of the hall).
Thanks to translucent walls, with a changing chromatic and luminous intensity, the atmosphere within the whole restaurant area is different throughout the day.

The staircase, composed of glass steps mounted on two steel blades, distributes the spaces of the various floors. The interiors have been entirely made of glass - chiseled on Fuksas's design - and transparent or translucent Plexiglas matching the satin stainless steel.
The window in Chater road is enriched by graphic markings made of blown glass neon tubes.

ARMANI THREE ON THE BUND
Shanghai China 2004

The plan for the Emporio Armani within a neo-classic building dating back to 1900 in the Bund, Shanghai's landmark area, is characterized by soft and seductive lines and volumes. The space is conceived as a monochrome, white volume. The monochrome shade confers a sense of infinity on the space, which thus loses its edges, gets expandable and conquers its bordering areas. The leitmotif of the Emporio Armani in Shanghai is a continuous steel and Plexiglas "track" that unfolds suspended through the space, with dress displays hanging from it. Cylindrical changing rooms with transparent acrylic finish on satin stainless steel metal sheets are organized at regular intervals on both sides.
The walls and the ceiling are pale gypsum made, the floor is pale blue painted with a gloss finish and all are combined together to create a dimension where everything seems floating. Such effect is further enriched by transparent acrylic horizontal shelves looking suspended in midair. Light sources dangle from the ceiling through a vertical forest of plexiglas shafts bringing the light downward. The entire store takes on a sense of fluidity that allows the creation of constantly new and different atmospheres, mirroring an ever-changing urban landscape, like that of Shanghai.

ARMANI Ginza Tower
Tokyo_ Japan_2007

It is always difficult to crystallize someone's image, and all the more so when it comes to Giorgio Armani, one of the most popular man in the world. In the Tokyo's flagship store, it was necessary to make his mood come to light alongside his own designer's creativity; it was about recreating the atmosphere of the famous Italian fashion designer's atelier as well as his aesthetics and image. In Tokyo, for the first time, the entirety of his production is contained within the same building. How to translate these elements into architecture? How to combine the idea of splendor with a sober elegance, the idea of modernity with the permanence of a style...... the Armani style? His image, the ceaseless research about material combined with luminous, transparent, delicate colours have been the elements from which our reflection has started. Our proposals have been multifarious. We have experimented new textures and shaped, sculpted, emptied, dematerialized spaces using light, the evanescence of an intimate perception born from the outside. The sophisticated image of Giorgio Armani's brand, full of transparencies and intimacy, is countered by the immediacy and contemporaneity of the spaces dedicated to the Emporio, where white light cuts cleave the ambient while reflecting in it.

The GA shops have been developed by offering the public "tiny drawing rooms" where various commodity groups are exposed. The "drawing rooms" are delimited by a carpet, clothes-hangers and a dividing panel, which is made by interposing a platinum coloured metal net between two clear glass layers. Floors and ceilings are characterized by a dark black colour and composed respectively reconstituted marble slabs tiles in Armani black and polished black steel plates.

Each partition of the shop has been made by using a sheer system where several satin and/or clear glass layers have been paired with a platinum coloured metallic net stratum which has been scanned in some points to create a bamboo wall effect.

The EA is characterized by continuous band walls, obtained through calandering, laser perforation and gloss black coating of steel plates.

The graphic shape of their cut derives from an original elaboration of the well-known "Emporio Armani" symbol, which comes into view with less or more intensity on every backlit shop wall. Through the bending of the continuous band, unique meanders have been obtained where the merchandise is displayed.

The restaurant recalls the motif of both the GA shops and the EA ones. The vertical partitions are attained through the same system characterizing the separating septums inside the GA shops: a platinum coloured metallic net between two clear glass layers. The ceilings and the floors are black as in the GA shops and feature the same properties. The black varnished wooden tables are adorned with a cavity containing a golden leaf under the surface and are surrounded by a series of "portals" made with steel layers in brushed gold finish, conceived as pierced shields according to a bamboo leaf-shaped motif.
The continuous band walls of the privé are backlit and calandering steel made, featuring a leaf design (like in the restaurant portals), the bar and retro counter are in black fiberglass and the floor is black as in the restaurant. It is basically formed by two areas: the internal one, having the same style and similar technical details of the restaurant, and the external one, a special teak-floored dehors enclosing a little garden with bamboo vases and grass level lights.

In the Ginza project the outside is a glass tower, perfectly harmonizing with the skyline. The lower part houses metal covered, golden coated, LED backlit bamboo trees. Luminous leaves with a variable intensity and colour depending on the moment of the day and season, are associated to these bamboos and they tenderly descend from the trees along their sides. On the upper floors, from the forth up to the eleventh, the façade no longer presents external elements; the bamboo design is carried on through a special curtain while the LED leaves are put behind the glasses. It has been like working under a microscope, surveying any possible little detail to find the right solution.

*

Armani
Fifth Avenue
New York
2009

Situated in midtown Manhattan in one of the world's best-known streets, the store occupies the first three floors of two buildings on 5th Avenue and 56th Street.

The four-level showroom (there is also a basement level) is a single, fluid, intercommunicating space held together by the impact of the staircase, the heart of the building, which connects the first (ground), second and third levels. Made of rolled calendar steel coated in plastic to enhance its sculptural effect, it is a totally free-standing structure that defies any simple geometrical description, a swirling vortex of bands that barely touch each floor as they glide over the vertical surface of the interior, thwarting any attempt to define their geometry and statics.

The general layout of each floor is an ever-changing pattern of curved surfaces that add visual interest to the light, putty-coloured walls.

Nothing remains unaffected by this interior movement, not even the external façade. Though aligned with the rigid Manhattan grid, it simulates the building's internal movement in images and colours projected onto a LED curtain. As well as being an external projection of internal space, the façade pays a special tribute to New York City, the yardstick of modernity and dynamism that all architects have to reckon with.

The internal space is made fluid by continuous wall bands faced with monochrome lacquered wood panels. The different radius of each wall curve create bends and twists where products can be displayed. In some places they accommodate fitting-rooms and VIP lounges; in others, staff areas, cash desks and special product locations such as Armani Dolci. Special attention has been paid to the lighting, which defines and emphasizes the curvature of the walls and spaces, and locates functions within the general layout.

The movement of the staircase is evident in every aspect of the interior design, from floor and wall display units to desks and armchairs, which mirrors, enhances and becomes part of the same vortex. The layout and circulation implied by the staircase are equally fluid. Glossy walls and furnishings are offset by black ceilings and black marble floors.

Similarly, the apparent simplicity of the interior is offset by the otherworldliness of the Armani/Ristorante heralded by the elevator entrance's curved bronze panels whose colours and reflections are foretaste of things to come. Sunken floor lights draw attention to the sensual curvature of the wall leading to restaurant, where colours and materials, though the same as in the rest of the store, are put to different use to create a more relaxed, recreational ambience. The restaurant has stunning views over 5th Avenue, Central Park and beyond, filtered through an amber haze.

Chairs, Tables, Benchs, Handles, Wall Units, Lamps, Porcelain Sculptures, Pavements, Coffee Machine, Chandelier, Carpets, Mirrors and a nice Jewellery Collections, the Design of Doriana and Massimiliano Fuksas fill the second part of this book.

auditorium chair BiBi for Poltrona Frau, 2005

The linearity of the future, that opens and closes like a box. It's the seat created for the Auditorium of the Grappa Nardini Research Center. It's a structure with harmonious lines, covered with the famous Pelle Frau, resulting from artisanal knowledge and modern technology and obtained from an exclusive craftsmanship process: 21 different leather treatments. The seat presents two options of use: it can be used "closed", almost as a seating pad, or "open" as a seat. The BiBi conceals an electronic device that can open the back of either one seat, one row or alternated seat, depending on the needs of the moment. Alternatively it can be opened and closed manually, as a classical theatre seat but with a revolutionary design: instead of lowering the seat, it is possible to open the back. The system of seating changes the rules, providing a direct fixing on the flat surface of the steps and allowing an easy access to the rows.

1

2

3

office chair
boa
for Luxy, 2006

Bea is inspired by the curvy female form and incorporates all the ergonomic technology within its flexible and sensual shell-shaped design. A new type of patented interface-user has been developed: "push & click" controls, lumbar adjustable settings, adjustable seat depth, adjustable armrests height and angle. Through these controls, and their many possible combinations, any single chair may assume more than 12,600 different layouts. Bea can be opened, closed, extended, retracted, lifted, bent to be adapted to any user's needs.

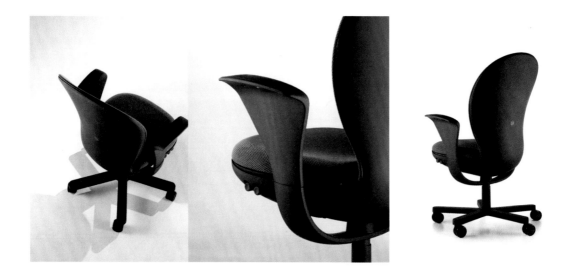

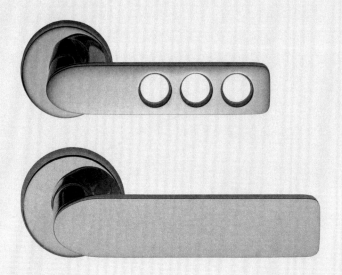

Blue Mountains
For Valli & Valli, 2002

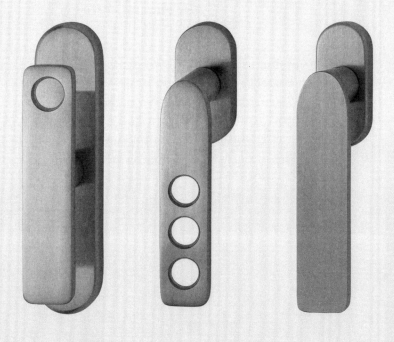

Carmen
For Manital, 2010

In its version for doors and windows, Carmen is manufactured in drop-forged brass with either chrome, satin chrome and satin nickel finish. The central fluting at the front and the back can be either red, white, black, yellow or copper.

Design awarded, "International Design Awards 2009-2010" Los Angeles, USA, 2011

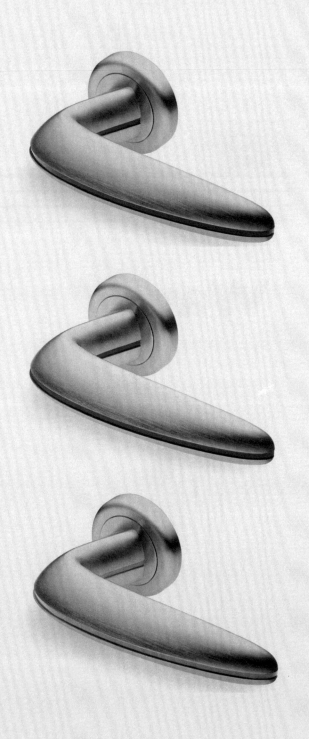

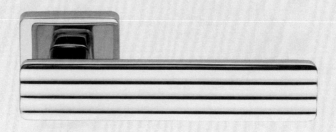

Wave

For Fusital, 2011

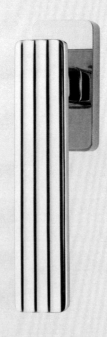

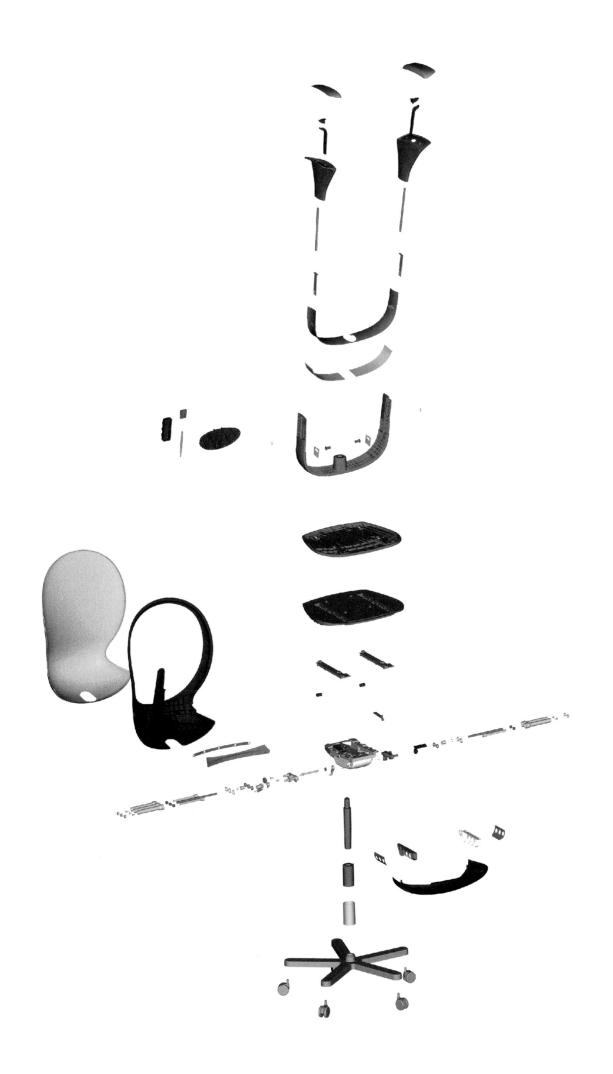

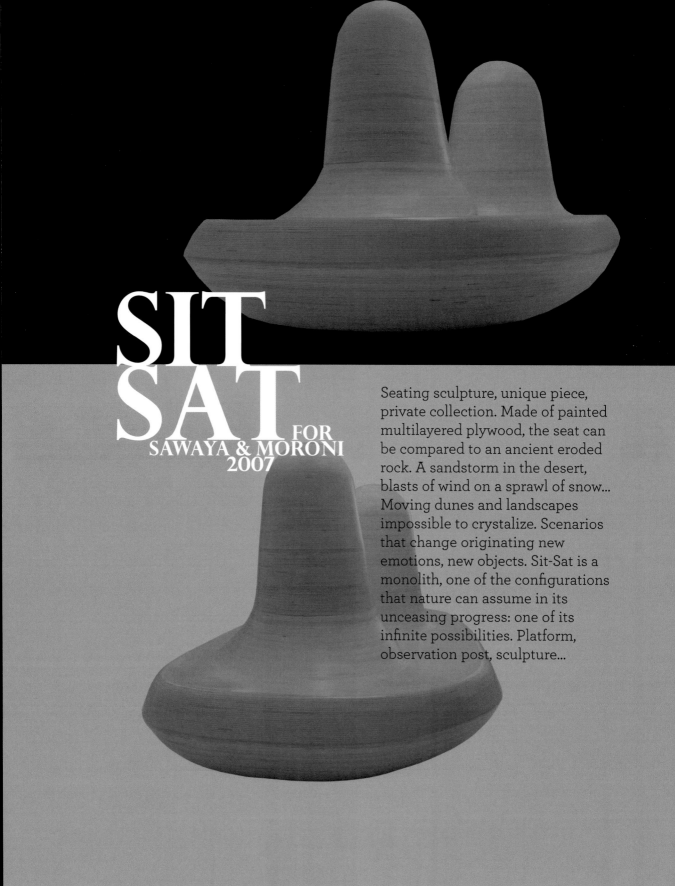

SIT SAT
FOR SAWAYA & MORONI 2007

Seating sculpture, unique piece, private collection. Made of painted multilayered plywood, the seat can be compared to an ancient eroded rock. A sandstorm in the desert, blasts of wind on a sprawl of snow... Moving dunes and landscapes impossible to crystalize. Scenarios that change originating new emotions, new objects. Sit-Sat is a monolith, one of the configurations that nature can assume in its unceasing progress: one of its infinite possibilities. Platform, observation post, sculpture...

CAROLINA
LOUNGE CHAIR FOR POLTRONA FRAU, 2008

A cosy nest, whose uninterrupted form combines set, backrest and armrest, that mixes the warmth of wood with the tactile charm of leather. A lounge-chair ideal for the relaxation areas of the home. Carolina has a comfortable curved shape: the ample armrests are an integral part of the shell and run along the enveloping seat and backrest. The leather upholstery for the supporting frame and pleasantly contrasts with the either natural or stained oak finish of the armrests. The cross stitching and buttons, also covered in leather, create an elegant decorative motif.

Auditorium Chair # Carla

For Poltrona Frau, 2012

The chair is formed of two planes that intersect and rotate their way into the back, chair and armrests, just like a flower. A measured, minimalist shape that can make the best use of even limited spaces.

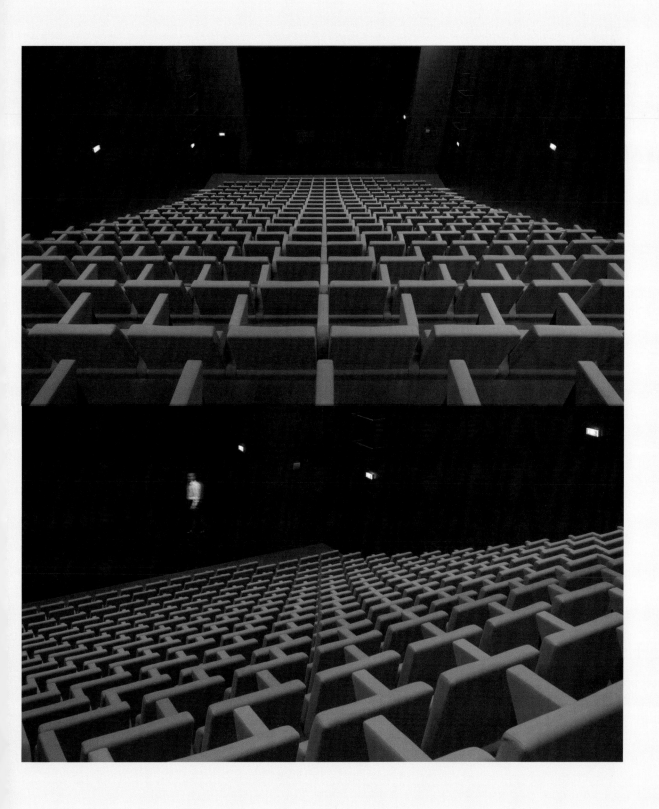

Dolly and Molly are two very different armchairs designed as ideal seats both for domestic and office spaces. More formal, composed, elegant with slender feet, Nabuck Rose leather upholstered, with Kashmir Nuage leather trim, "Dolly" has been designed in various colors, ranging from pastel to brighter hues, to create a "classic" yet contemporary piece of furniture. Big, rounded, curvy, comfortable and important, white Kalgan fur upholstered, "Molly" welcomes its guests with a long hug, holding them in the comfort of its shape. Ideal for reading, contemplating and relaxing, "Molly" is not just a piece of furniture, but it's like a sculpture, the ideal complement to large spaces, lounges and all those areas that need strong points of attraction… Armchairs with a different design, whose shape merges with the warmth of their wood frame, covered by the tactile charm of leather and fur.

Molly

Armchair for Baxter, 2013

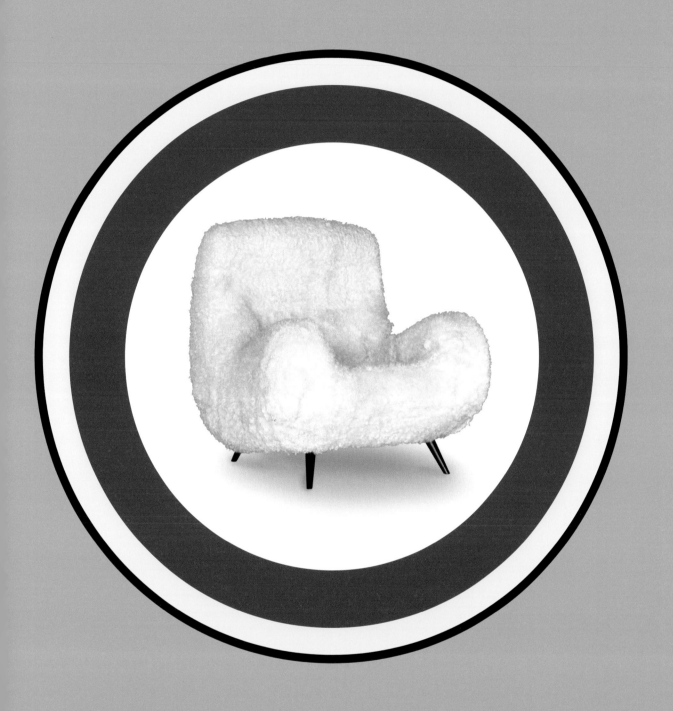

Dolly

Armchair for Baxter, 2013

TABLE AND BENCH
TOMMASO

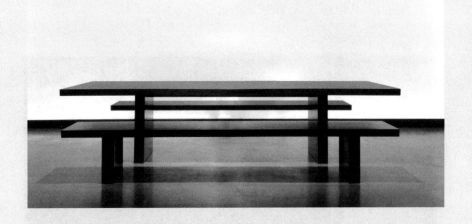

Table. Detachable frame. Top thickness 6 cm and legs 50/55 x 10 cm made of bent natural sheet steel.
Bench. Top thickness 6 cm and legs 34 x 10 cm in bent natural sheet steel.

FOR ZEUS-NOTO, 2009

WALL UNIT
NINA

Three size wall unit. Bent sheet steel frame, hollow box construction. Rust or aluminum epoxy coated inner box. Vertical or horizontal fitting.

FOR ZEUS-NOTO, 2011

STACKABLE CHAIR

BIANCA

& BAR-STOOL BIANCO
FOR ZEUS-NOTO, 2004

Stackable chair and Bar-Stool. 15 x 20 mm sandblasted stainless steel frame in either gunmetal epoxy coating or semi-opaque white. 10 mm opaline/glossy white or opaque/glossy black pressed acrylic resin seat.

LE EMOZIONI NON VANNO RACCONTATE, VANNO VISSUTE.

Milano, via Medici 15 · Roma, Via Gregorio VII 308/310 | www.baxter.it

MADE IN ITALY

LAMP lavinia
For iGuzzini, 2003

The Lavinia system, which was conceived for exterior lighting, features a single, double and triple optical opening that can be adapted to the slope of the street surface and can be installed either on poles or as single -, double- or triple-arm wall lamps. According to the lighting sources, it can be used for lighting low-traffic streets, residential quarters and large areas.

Nuovo Polo di Fiera Milano, Rho-Pero, Milano. Progettisti dell'architettura e della regia luminosa: Massimiliano Fuksas ③ e Doriana Mandrelli ②. Committenti: Fondazione Fiera Milano e Sviluppo Sistema Fiera, Presidente Luigi Roth ④. Consulente tecnico per l'illuminazione: iGuzzini ①. Apparecchi iGuzzini utilizzati: Lavinia, design Massimiliano Fuksas e Doriana Mandrelli.

New Milan Trade Fair, Rho-Pero, Milan, Italy, 2005

Minimalist clean shape for a LED system designed to illuminate residential areas and for interiors.

Lamp Zyl for iGuzzini, 2010

Chandelier Moony for La Murrina, 2011

One, ten, twenty, one hundred composed pieces forming a white, transparent, shiny or satin-finish lamp in Murano glass, and letters and symbols of an archaic alphabet writing the light...

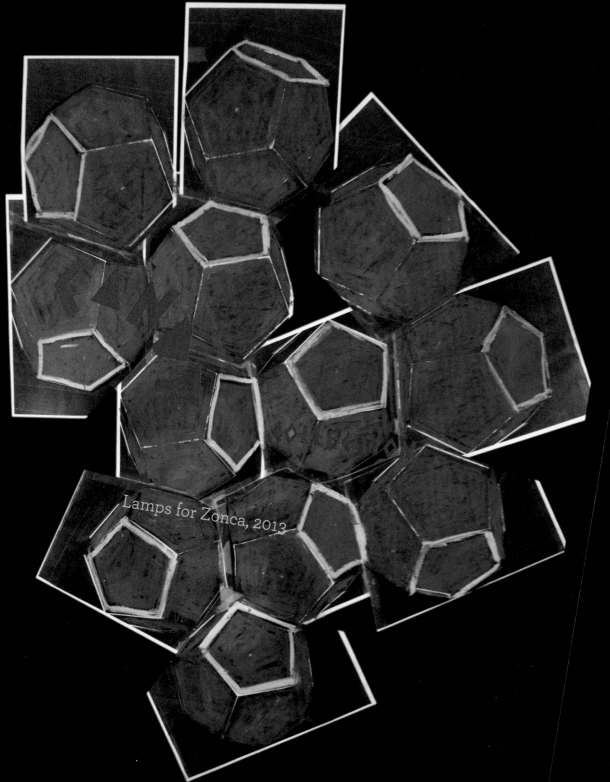
Lamps for Zonca, 2013

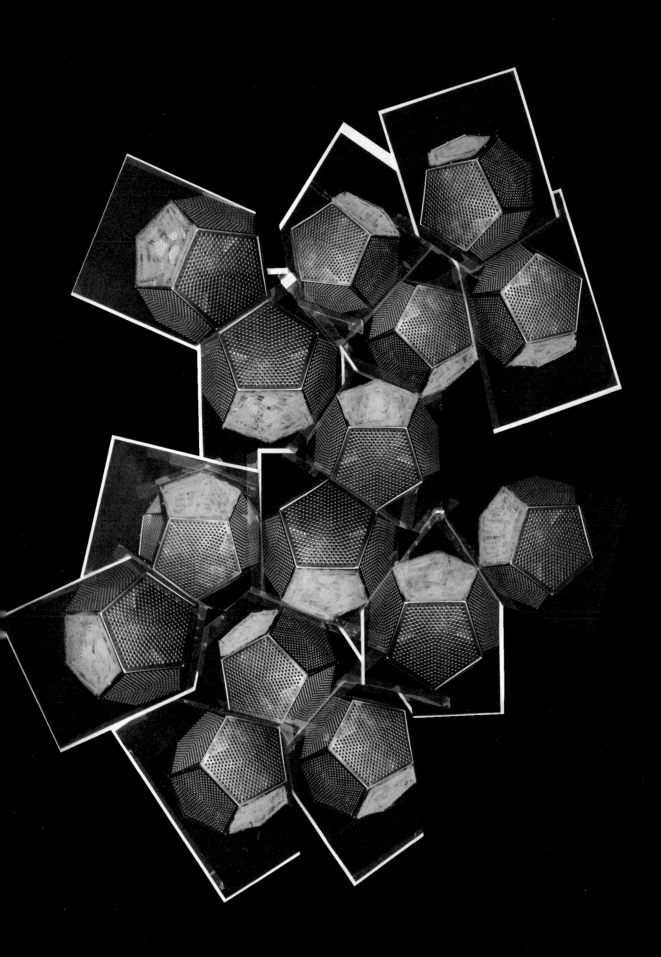

Lamps as coloured candies stuffed with light. The collection of lamps made for Zonca originates from a geometric design: a shrine made up of twelve pentagonal sides composing the framework of each lamp. The solids and voids effect is obtained from a body conceived as a frame which remains visible and a microporous covering for the sides. Whether they be used as floor, pendant or table lamps, the idea is to play with each lamp in order to create a unique ambiance. With such a collection of coloured lamps, conceived to be connected, overlapped and inlaid to generate an always original lamp, it is possible to create an installation, a sculpture or a luminous path. The collection is also equipped with a stylized pentagonal prism, which can be inlaid into any lamp, to direct the light beam.

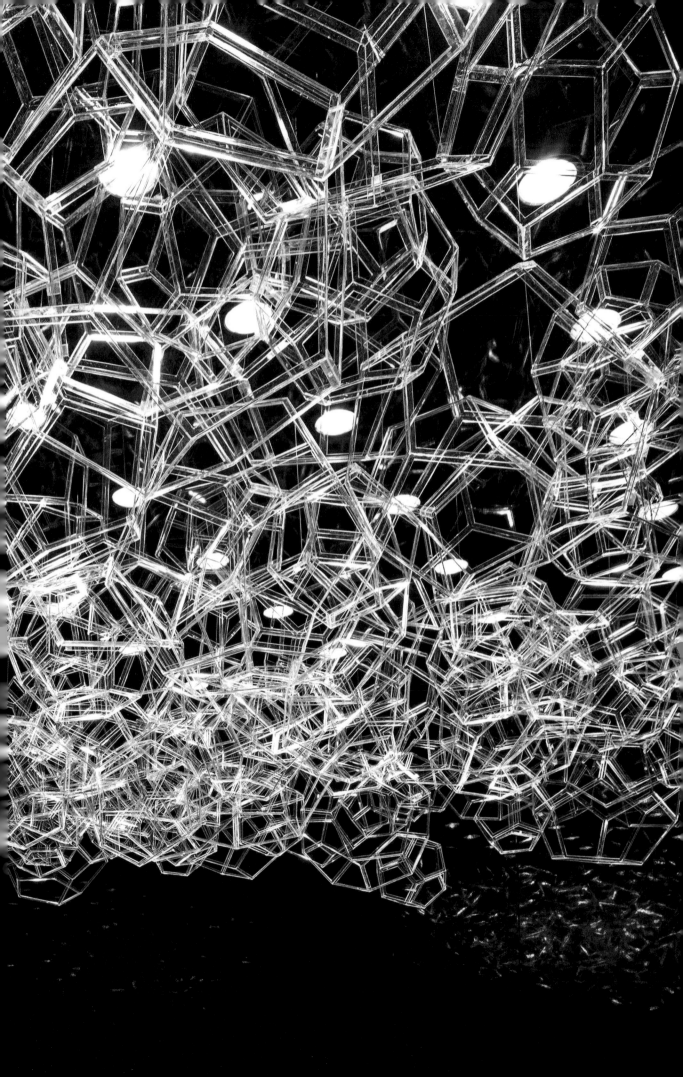

The Impronta collection was born as part of the project "Catalano per l'Architettura". In contrast with more rational frameworks and geometries, the organic matrix becomes the collection's central thread. An 80x45 cm built-in washbasin, a 120x80 cm shower tray and a backlit mirror are additions to the collection's progenitor, a 125x50 cm washbowl, integrated with a unique, originally- designed towel rack and the shiny steel chest of drawers with chrome aluminum bases. The shower tray, framed by a regular quadrangular perimeter, evokes a delicate natural element: the original anti-slip motif is made in the same way as the sea- waves-shaped sandy surfaces delimiting sinuous bottomlands where water flows. A superimposition of ceramic baseboards forms the washbowl as a basin nearly carved in matter, while the mirror converts the original concept into a clever game of transparencies and light. Original is the side view exhaust system of washbowls, that have been devised to house either one hole, three hole and wall faucets.

WASHBASIN, SHOWER AND FURNITURE

IMPRONTA

FOR CATALANO, 2011

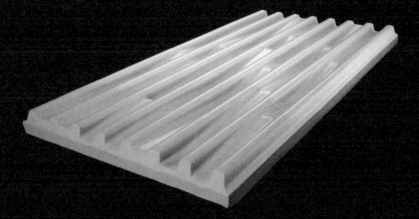

Dimensions: 20 x 40 cm
A porcelain stoneware sculpture that "catches" and sends back the natural light as the day goes by.... A ceramic fragment devised either as an architectural sculpture or as the rendering of an idea of colour rather than a form for interior design.

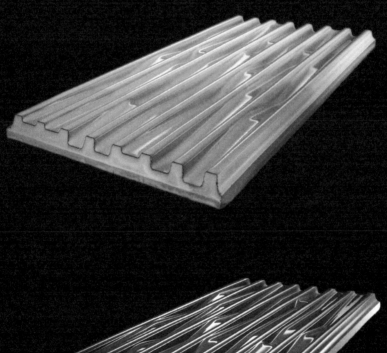

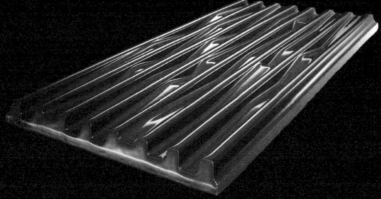

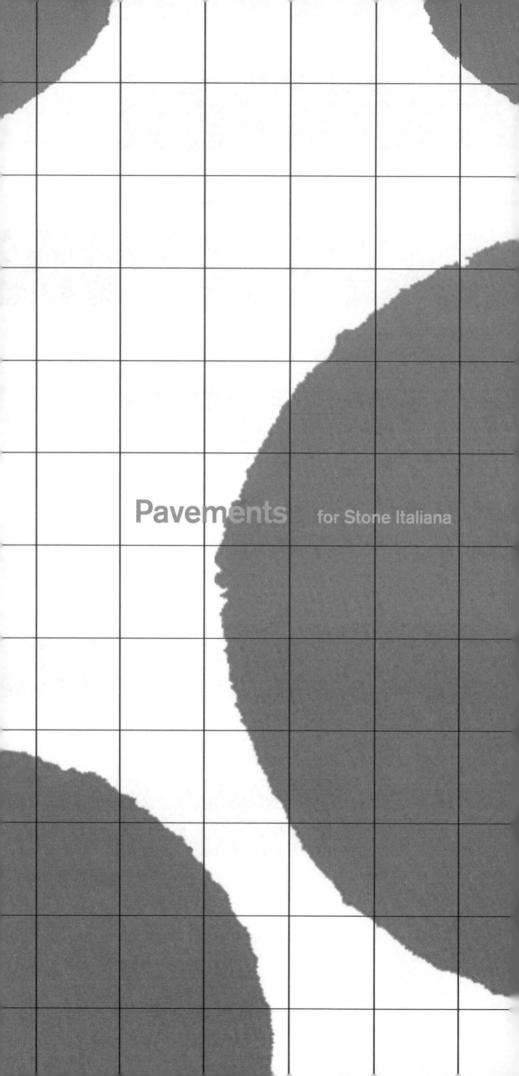
Pavements for Stone Italiana

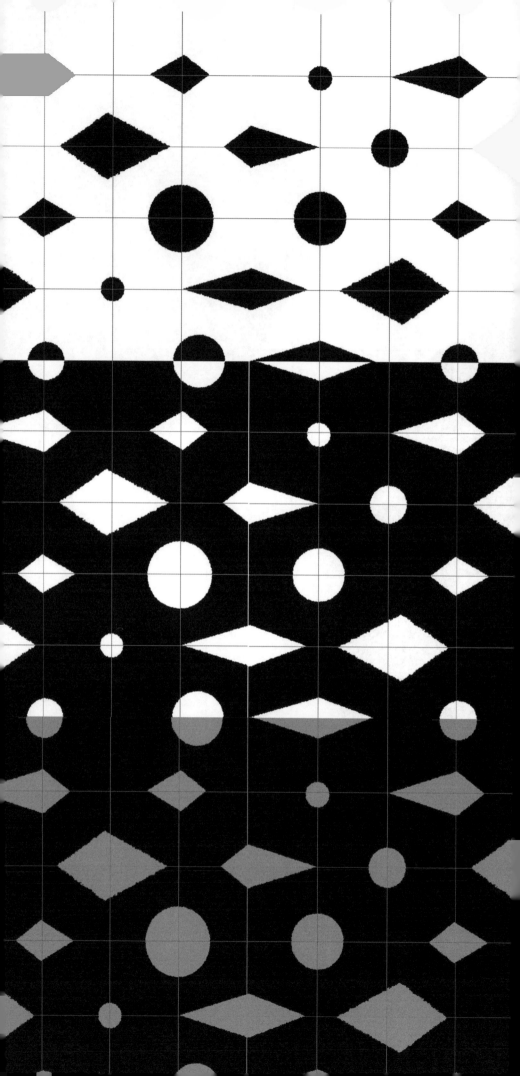

A large carpet, entirely woven on manual loom by skilled artisans, composed of sewn patches symbolizing the union between design and tradition.

"Tappeto Sardo" Sardinian Craftsmanship and Design
Salone Internazionale del Mobile, Milan, Italy, 2006

Silver Set
For Sawaya & Moroni, 2007

Carafe and glasses, collection of handmade sterling silver. The pieces lie down on a long shining glove-box that reflects and multiplies the sinuous necks of the carafes, bloomed from a softly flared figure, and the cone-shaped glasses moulded by caressing hands. The primary shapes, almost like those of artist Giorgio Morandi, are softened by the luminosity of the materials. The wide and flat surface seems to be designed to welcome the lingering reflections made by the skillful manufacturing.

MUMBAI

TABLE AND LIBRARY FOR SENIOR MANAGEMMENT FOR HAWORTH CASTELLI, 2008

The surface of Mumbai bends and redoubles, thus creating a blank space that becomes the centre of the project. A plywood sheet, folded up to the limits of the material's mechanical features, defines and encloses a lattice structure made of laser cut steel. A light and discreet architectural "heart" that supports a thin shell. An innovative exploitation of a traditional material. Sophisticated folding technologies emphasize the characteristics of preciously veneered plywood, and enhance the lightness of the shape.

A large table is the main element of the range. 2 veneered plywood sheets, 14 mm thick, folded with high frequency press technology, define the perimeter shape of Mumbai desk and create and inner empty space housing the metal bearing structure. The range can be enriched with a storage unite of different size: a high cabinet or a lower one. The external envelope of the cabinets is made of one folded plywood sheet outlining the shape and the bases of the cupboards. Back, vertical dividers and shelves made of laminated plywood with opaque black finishing complete and stiffen the cabinets. Both storage unit typologies can be equipped with 2 side hinged doors made of plywood.

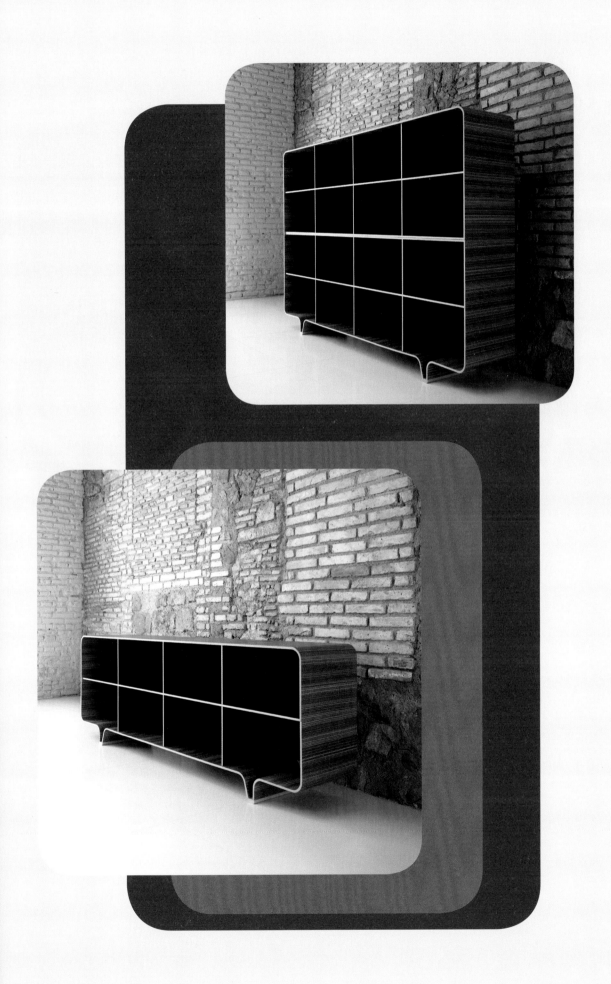

Rosy
Mirror for Fiam, 2013

Mirror: reflection, reversal, doubling. Two wall mirrors conceived as unique pieces, entirely made of glass. "Rosy" is a single rotund body with rounded frame, resembling an eclipse of the moon, or perhaps a pool set in a lunar landscape. Its companion, "Lucy", is a sculpture with an elliptical reflective surface at its core, set in a rounded frame shaped like three large, irregular petals. Both silvered glass frames have sensual, feminine lines. Mirrors that add character to a design scheme, each one a piece in a jigsaw puzzle or in a map waiting to be put together. Mirrors to look at yourself in, for pure vanity, or in search of something that is legible only when reflected in them.

Lucy
Mirror for Fiam, 2013

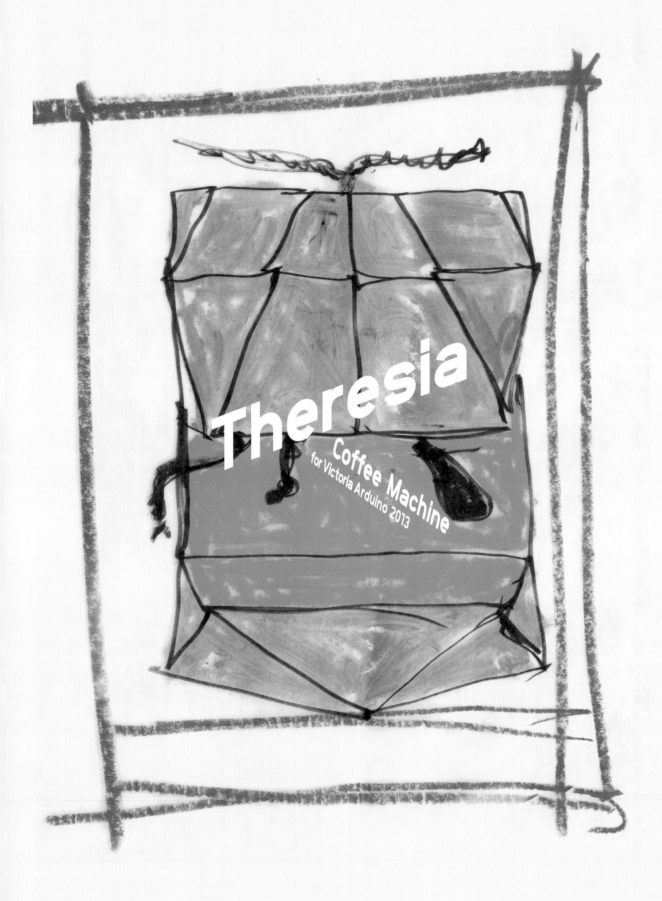

The espresso coffee machine present itself in a compact volume, as a sculptural object with a geometric design. The supermirror stainless body has been envisaged as a hollow diamond . A precious tool with a charming design, and at the same time simple and functional. A machine allowing ourselves the daily act of enjoying an espresso not only as a pleasant break, but also as an authentic little luxury.

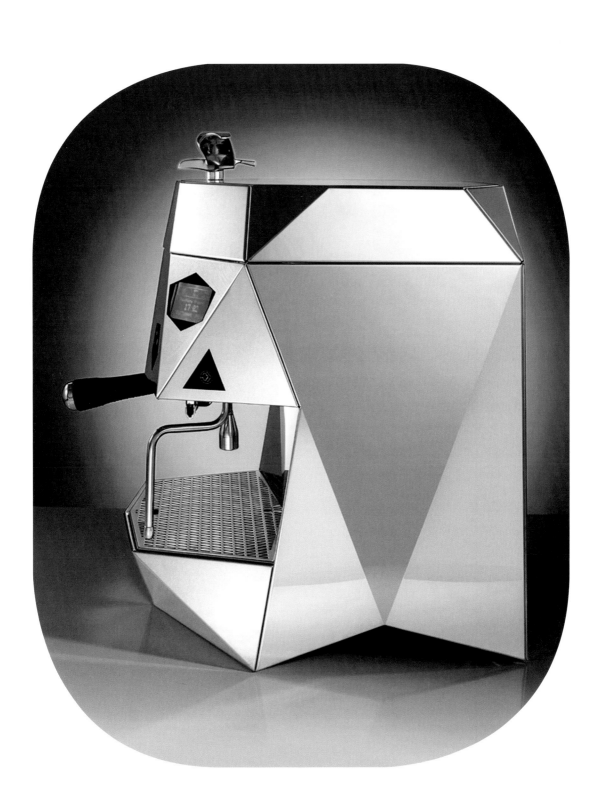

Gipsy
Jewelry collection
for Sicis, 2013

Objects found by chance, jewels, amulets, memories of distant houses..... each one is a memory, a moment, an affection, secret stories to care for and carry with ourselves, golden jewels made of small glass and precious stones tesseræ.

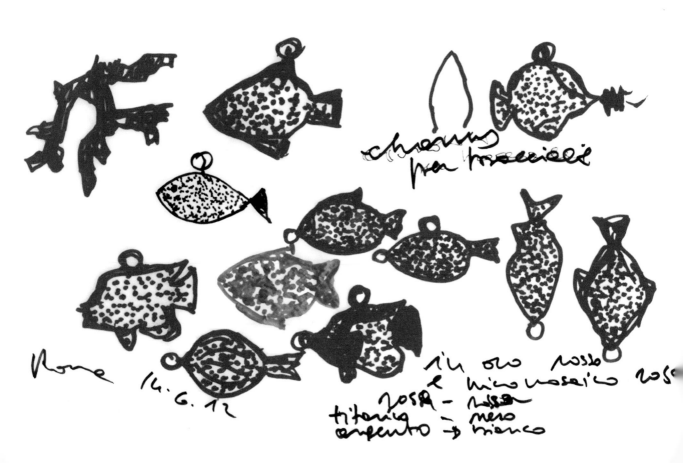

ISLANDS

A collection of furniture and contemporary jewellery, each in a limited production of 9 pieces, designed by the architects Doriana and Massimiliano Fuksas together with the artist Mimmo Paladino. For Short Stories, 2006

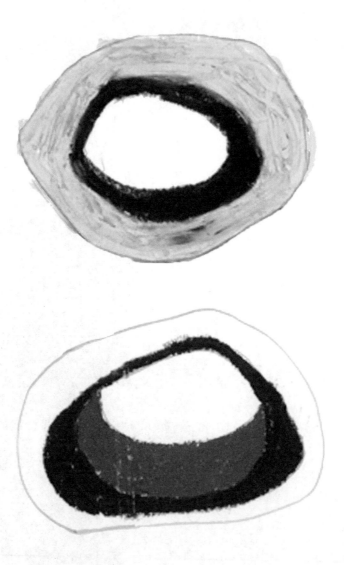

ISLANDS. Set of jewelry pieces – sculptures in silver. Comprising necklace, bracelet, earrings and ring. Limited edition of 99 signed pieces.

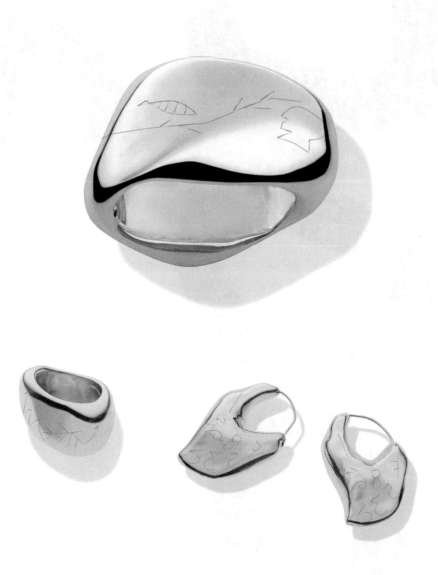

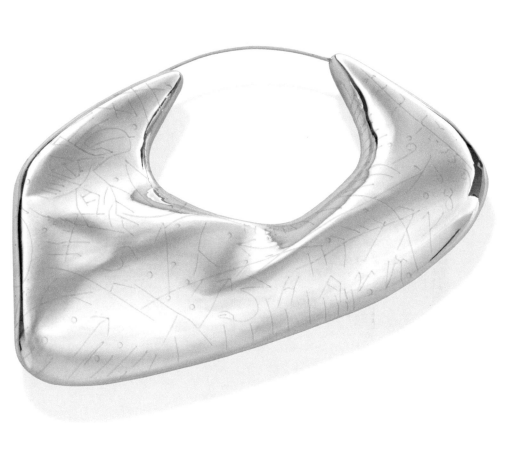

ISLANDS. Set of jewelry pieces – sculptures in 24 k gold. Comprising necklace, bracelet, earrings and ring. Limited edition of 9 signed pieces.

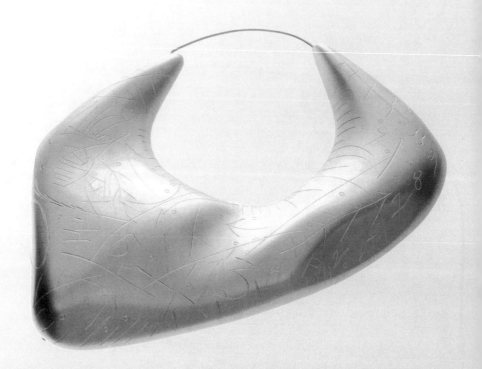

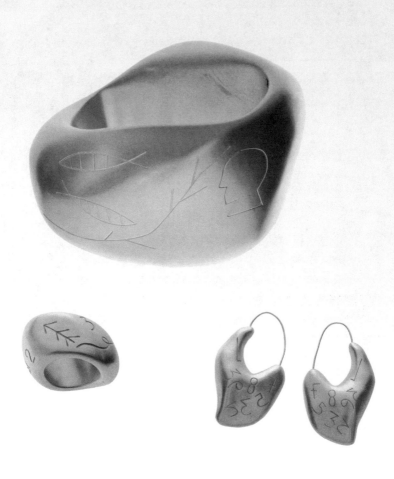

MEFLECTO. A work comprising 4 cast aluminium wall-panels with folding legs to let the panels double as tables. Limited edition of 9 signed pieces.

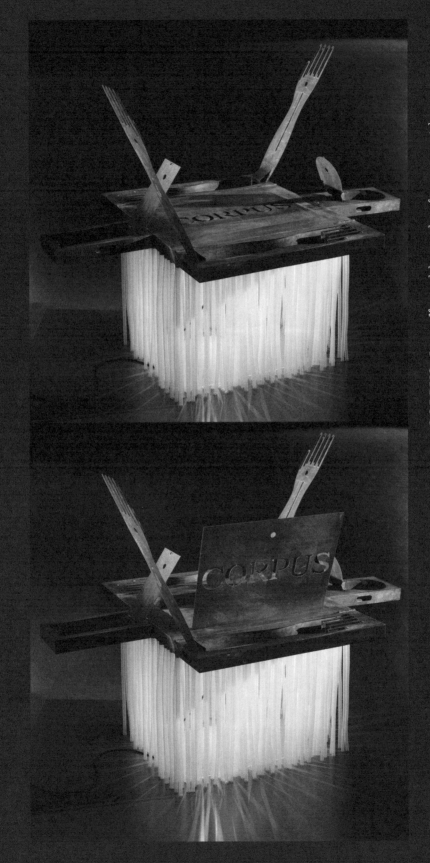

CORPUS. Lit coffee table made of cooper with silicon legs. Limited edition of 9 signed pieces.

ARCHITETTI ILLUMINATI E ARTISTA
Silver candelabra. Limited edition of 9 signed pieces

GLI ARCHITETTI ILLUMINATI
LUCE DELLA RAGIONE: LA
SAPIENZA DELLE ZAMPE
DIRITTE CURVE SGHEMBE DEI
TAVOLI MASSIMILIANO DORIANA
UN NOME LUNGO SCANDITO
DAL VORTICE DI UNA NUVOLA
DALLA LEVIGATURA DI UN
SASSO ISTORIATO: UN PITTORE
CHE DIPINGE "ORO" DI ROSSO
LE PEPITE, COME CARDINI
PREZIOSI: L'ALLEGRA MEDUSA
FOSFORESCENTE DEL "CORPUS"

M. M. S'...
15 LUGLIO
2012.

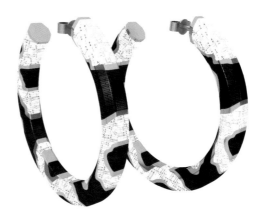

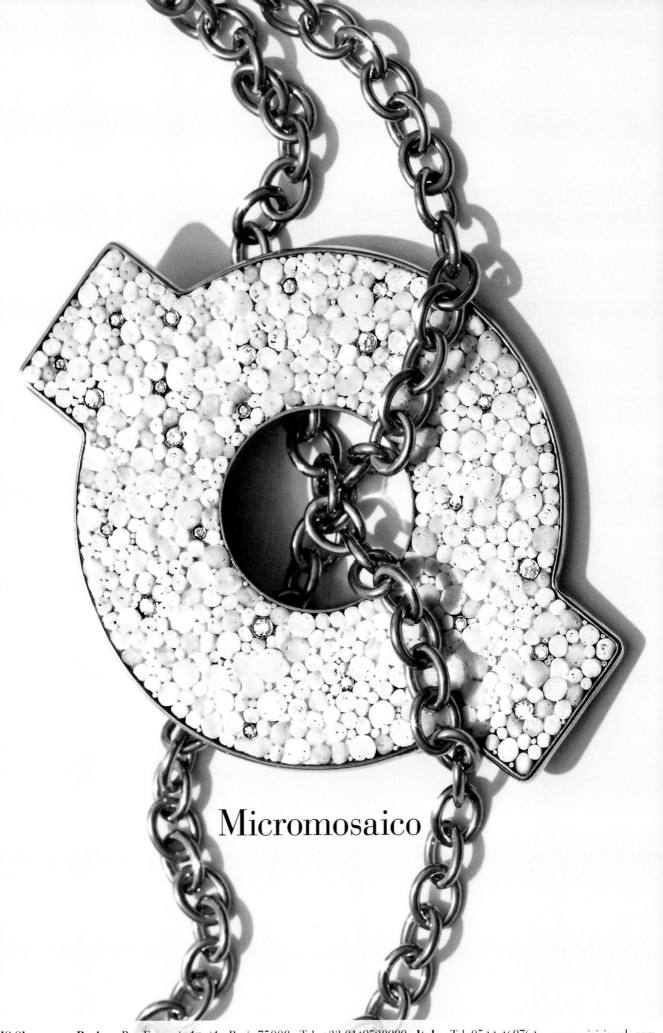

Micromosaico

CIS Showroom Paris · Rue François 1ᵉʳ, 41 · Paris 75008 · Tel: +33 0149528989 · **Italy** · Tel: 0544 469764 · www.sicisjewels.com

Creating atmospheres,
a combination of objects, colors,
materials and sensations.
The <u>interior design</u> production
by Doriana and Massimiliano
Fuksas represent a natural
extension of their architectural
process of creating spaces.

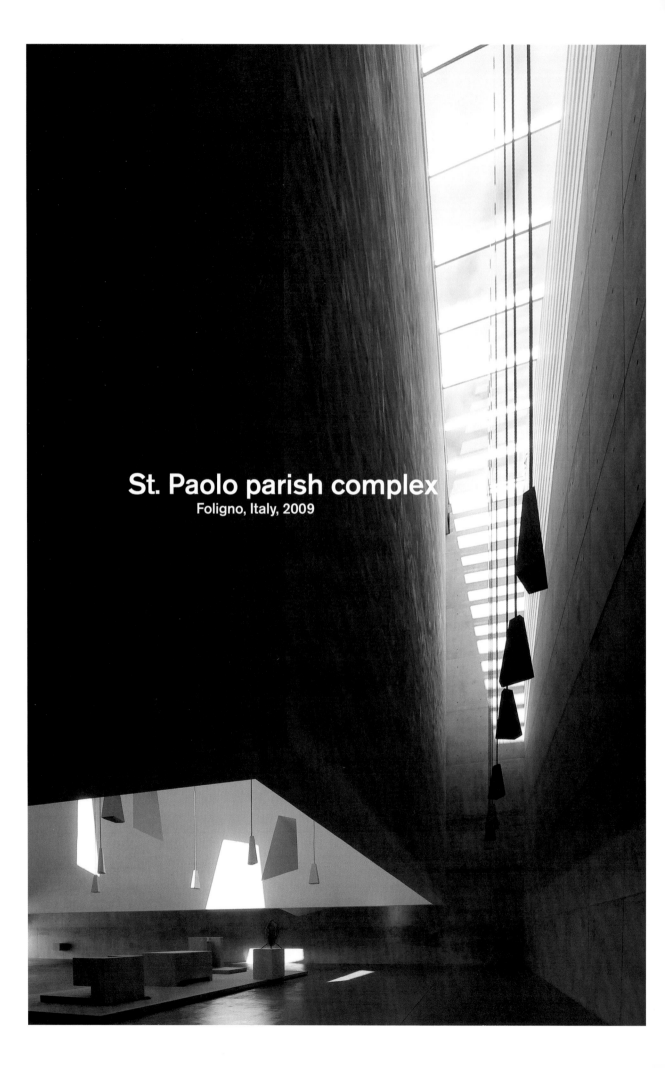

St. Paolo parish complex
Foligno, Italy, 2009

The interior was intended to emphasize the centrality of the Altar, whereas the Baptismal Font and the Ambon are located asymmetrically to the former.
The dominant idea is to underline the active role of the celebrating Assembly.
The design of the interior and the lighting bodies evokes the idea of essentiality and purity. The sober shape of the oak pews inspires meditation. The religious ornaments made of stone, such as the Altar, the Ambon and the Baptismal Font, are emphasized by the beams of light crossing the monolith both vertically and obliquely. The effect created during the day by the natural light is recreated at night by hanging lights whose angular shape retraces the profile of the openings carved on both volumes.
The Master Enzo Cucchi has created the monument "Stele-Croce" for the area outside the Church, a 13,50 meter high sculpture, made of concrete, that also becomes an architectural element. The Master Mimmo Paladino has created the 14 iron sculptures representing the Stations of the Cross.

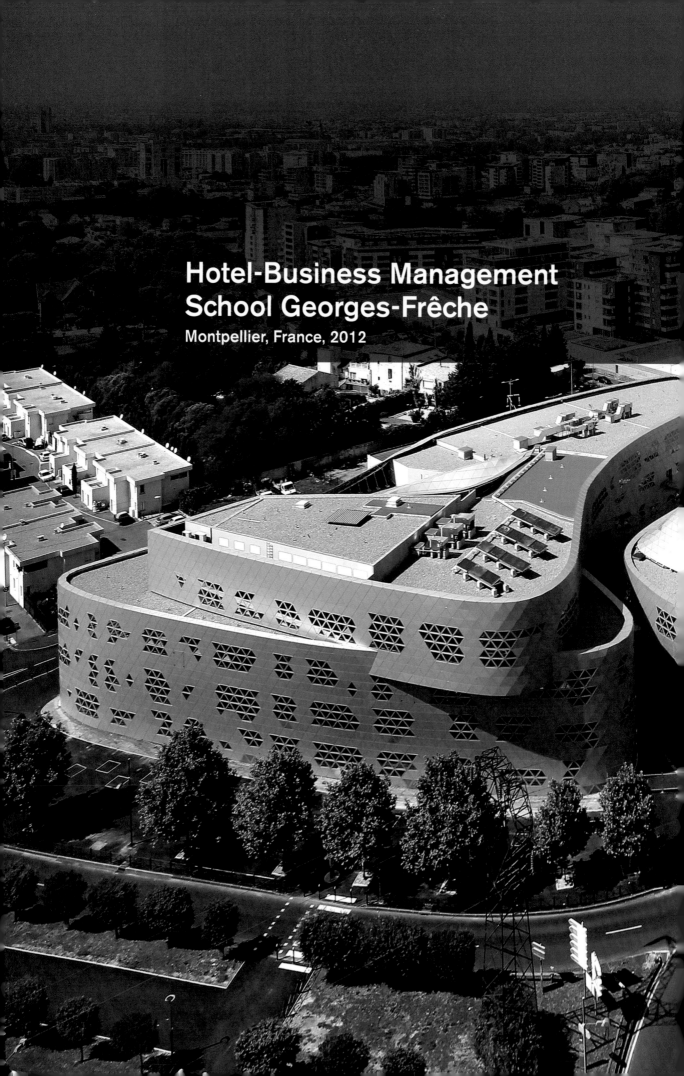

Hotel-Business Management School Georges-Frêche
Montpellier, France, 2012

The project, which is developed horizontally, comes across as a single entity. It has a formal diversity, compact volumes and sculptural shapes. The Lyceum is housed in two buildings making up the sculptural mass around which the gym, the students' residence and the management's housing gravitate. One building has three floors and includes: a multi-purpose room, an exhibition gallery, administrative offices, classrooms and a canteen with exits leading to the recreational areas outside. The other building - with two floors and a distinctive Y shape – hosts the activities of the students as well as the areas used for catering and hospitality.
A hotel - open to the public – with twelve rooms (6 of them are 2 or 3 star rooms, 4 are 4 star and two suites); 3 restaurants: a gastronomic one (with 50 seats), a brasserie and one for students' vocational training with an overall capacity of approximately 200 seats.
Fuksas architects have designed the interior, the areas accessible to the public and the spaces for the gastronomic sector and the hotel. The entrance hall, leading into the gastronomic

restaurant and the hotel, houses a reception desk: a white lacquered sculptural object, mirroring the fluid forms and the solid character of the structure. The desk is covered with materials that are used for making boat hulls. Different types of originally designed tables and chairs define the spaces dedicated to the interaction between the public and the students. There is also the limited edition furniture specially made for the hotel. The School walls and those of the students' residence are painted in a different colour on each floor, with shades ranging from yellow to green, magenta and orange. The colours serve as signage to distinguish the different spaces and activities. The facades of the building are made of 17.000 triangular-shaped small boxes in anodized aluminum, each one being a unique piece. The geometric design of the aluminum "skin" maintains continuity with the one of the all different, triangular-shaped 5.000 glass windows. The interaction among the facades cladding the architecture of the volumes highlights the solids and voids effect as well as the light and shade one characterizing the idea of the plan.

The project is composed of two main "bodies": one that extends horizontally, "suspended, lightweight, transparent"; the other with a tension in height, "anchored to the ground, imposing, reflective".

The first, stretching out towards the city, consists of cantilevered volumes called "satellites" that accommodate offices, the conference room and the exhibition hall. The facades, mostly glazed, give lightness and transparency to volumes of different proportions, that follow each other and overlap in "suspension" on the surfaces of the water.

The building that accommodates the Archives is an imposing monolith thought of as a place dedicated to memory and research. It houses the archival documents and the reading room. The facades of the monolith are coated with aluminium "skin" that runs throughout the volume, except for some glazed insertions that allow the amount of natural light in the reading room and the entry route. The "noble" sculptural building, with a basin in part lapped against it, evokes the idea of a precious object, a treasure chest, that is reflected in the water veil.

The basins insert themselves between the building of the Archives, the "satellite" volumes and at the foot of the satellite volumes. Walkways above them create a connection both between the cantilevered volumes and the two "bodies". The water veil becomes a vehicle of change for the architecture, designing voids and new spaces, thanks to the reflections and the play of natural light created by the cuts of the suspended volumes and the "skin" of the monolith.

The facades of both "bodies" follow a lozenge geometry that is repeated both in the aluminium cladding of the building of the Archives and in the glass facades of the "satellite" volumes. Between the monolith and the "satellite" volumes stands the artwork "Cloud Chain" by Antony Gormley.

The connection with memory is symbolically recognizable in Pascal Convert's work, a series of concrete "safe-boxes" set in the area opposite the "satellite" volumes, which portray bas relief faces of some personalities who have left a mark in collective memory. It is an artistic installation, firmly fastened to the ground as well as the volume of the monolith, that is reminiscent of roots immersed in the depth of memory.

A double-height hall welcomes the visitor. The "suspended" effect of the "satellite" volumes is highlighted by the art intervention by Susanna Fritscher which, through a minimalist touch that consists in the realization of false ceilings as stainless steel "sheets" shaded in red, emphasizes the interaction between the architecture of the complex and the lines of the "satellites" volumes. The red colour gives depth to the volumes that stand out horizontally at different heights, creating at the same time a play of solids and voids, between material and immaterial.

The entrance leads to areas dedicated to the public: the reading room, the exhibition room and the conference room. The interiors are characterized by large bays offering an overall view that gives the immediate perception of the importance and uniqueness of the place.

Refurbishment of the Ex Unione Militare Building

Rome_Italy_2013

A new architecture redefines the urban landscape of Rome historic centre according to contemporary taste. It's the steel and glass «Lantern» of the «Ex Unione Militare» building, situated between Via del Corso and Via Tomacelli, that crosses the four floors of the building from the ground-floor up to the panoramic terrace with a view of the dome of the Basilica of Saint Ambrose and Carlo al Corso.

A contemporary interpretation of the historic centre has led to a minor intervention on the outside of the building, whose original construction dates back to the end of the XIX century, focusing on the renovation of the interiors and roofing.
The restoration of the outside has focused on the recovering and valorization of the original architectural features of the building. The architecture of the facades has been highlighted through a minimal light design intervention, which gives a touch of contemporary to the building whilst putting it in connection to the city.

The large «Lantern» is the symbol of the project, the heart of the intervention.
A steel and glass triangular-shaped structure that crosses the entire building and contains the vertical connections, the service and accessory rooms as well as part of the plants. The full-height void created by the "Lantern" generates a glimpse along the structure of the various floors, which are interconnected through gangways.
The part of the «Lantern» functioning as roofing reaches a maximum height of 7,50 metres from the floor and accomodates a 300-square-metre panoramic restaurant space. From the panoramic viewpoints of the city, the «Lantern» hit by natural light takes on the appearance of an irregular mirror rather than an icy lake, whereas at night it lights up and takes on the shape of looks as a large lamp.

In the evening, the facades of the building light up and look like a theatre set, the atmospheres change and the audience can enjoy from the outside the internal show of lights and colours.

Each floor offers visitors a unique space characterized by white floors decorated with «bubbles» of different size and colour, ranging from red shades to orange and violet tones. The ground floor, set up as a large bazar full of colours, objects and accessories is a wide and porous open space connecting via Tomacelli to the adjacent square.

The interiors have been designed by Fuksas as shiny elements inspired by children toys. They are sculptural objects with fluid shapes mostly made of fiberglass and characterized by a shiny white colour. Tables, desks, pouf, display racks, «spinning top» for garments and accessories that gently fit into the exhibition areas.
The «spinning top» fan out like petals like artistic installations. Oval-shaped mirrors reflect the interiors, lights and colours as a whole. The «Lantern» acts as a theatre wing that opens on each floor. The false ceilings light up through a range of colours recalling those of the floors. A stair of led-light-lit glass steps, with shiny surfaces, connects and embellishes the spaces.

Duomo Tube Station

Naples_Italy_2005 – ongoing

Stazione Duomo is a stop in Naples' underground Linea 1 and is situated in one of the most well-travelled areas of the city – Piazza Nicola Amore, an historic neapolitan square at the junction of Via Duomo and Corso Umberto I - in terms of residents and tourists. Excavation works for the station have brought to light the foundations of a temple dating back to the I century b.C, as well as several archeological finds of great interest. The archeological find is crucial to the scheme of the station and the intervention aims to turn the temple into a unique museum area. In order to preserve the temple, a geodetic bubble-like structure with a steel and glass triangular frame has been devised to let direct natural light in. Glass allows visual legibility and continuity from the outside towards the inside, while also lending a touch of lightness to the volume at the centre of the square. The architectural addition of the glassy roofing "deforms" the urban axis and channels the attention on the square and the temple. The first basement level – conceptually inscribed within the base ellipse of the "bubble" – houses the temple. The second basement level of the station corresponds to the mezzanine floor. In this part of the station tied to mobility and urban journey, the perceptual experience undergoes a change. The rhythm of the journey is given by colours and geometric textures. The entrance to the underground is characterized by colours, recurring geometric motifs and reflective surfaces.

IS MOLAS GOLF RESORT AND HOTEL

PULA_CAGLIARI_ITALY_2006_ongoing

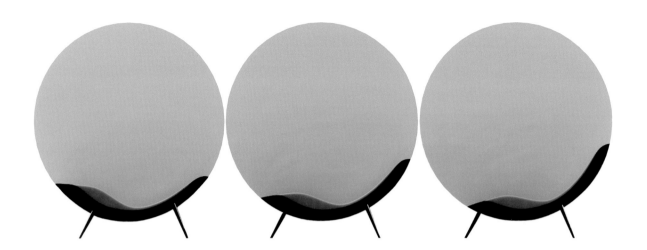

The plan for the new Is Molas Resort complex as a development of the existing Golf Club regards the building of three greens, two five star hotels, a number of villas, a spa and a club house. The idea behind the intervention on Is Molas complex - from the large master plan scale to the mansion's small one and to that of the most representative building, the club house - consists in a progressive aggregation of building units. It is reminiscent of fragments, of the archeological memories of Nuragic settlements. The deluxe hotel, set on the visual axis to the Nora archeological area, gives shape to the whole intervention. The entrance to the building materializes as a long yet pleasant path, a larch gangway over a pool of water. Once arrived to the imposing entrance hall, we find on the left the main restaurant with a broad external terrace, and on the right another terrace with a swimming pool; a system of foldable, fully opening and transparent glass walls allow the total fusion of the spaces in case of necessity. The hall, set at the end of the path, is a space opening towards the horizon and leading to the 80 rooms placed on opposite sides which are reachable by walking paths passing through verdant gardens of water and Mediterranean greenery.

Set on the upstream area of the hotel, at a higher altitude, the mansions are built as small villas with communal green spaces. Villas are made with natural materials produced locally. The treatment of the surfaces has been envisaged with the purpose of making them as similar as possible to the surrounding soils and stones, by using "raw" plasters, natural pigments and claddings consisting of stone slabs and local stone chip flooring which recall the typical Nuragic architecture.

The gradual "naturalization" of the plan finds its highest expression on the area housing the unit Spa – Hotel Suite; here the alignment sought after inside the central hotel is abandoned in order to open the view over the greens and follow the difference in height of the ground. Within the units of the Spa, a sophisticated trick of skylights on the vault roofs allows zenithal natural lighting in whilst creating a dense light and shade atmosphere. The Hotel Suite pursue the same expressive purpose of the adjacent Spa in the communal areas. The building then soon winds up to follow the higher slope of the ground. The rooms fan out and lie on two levels.

Placed on a higher barycentric position than the new greens', the golf structure finds in the new Club house the main junction overlooking the entire layout.

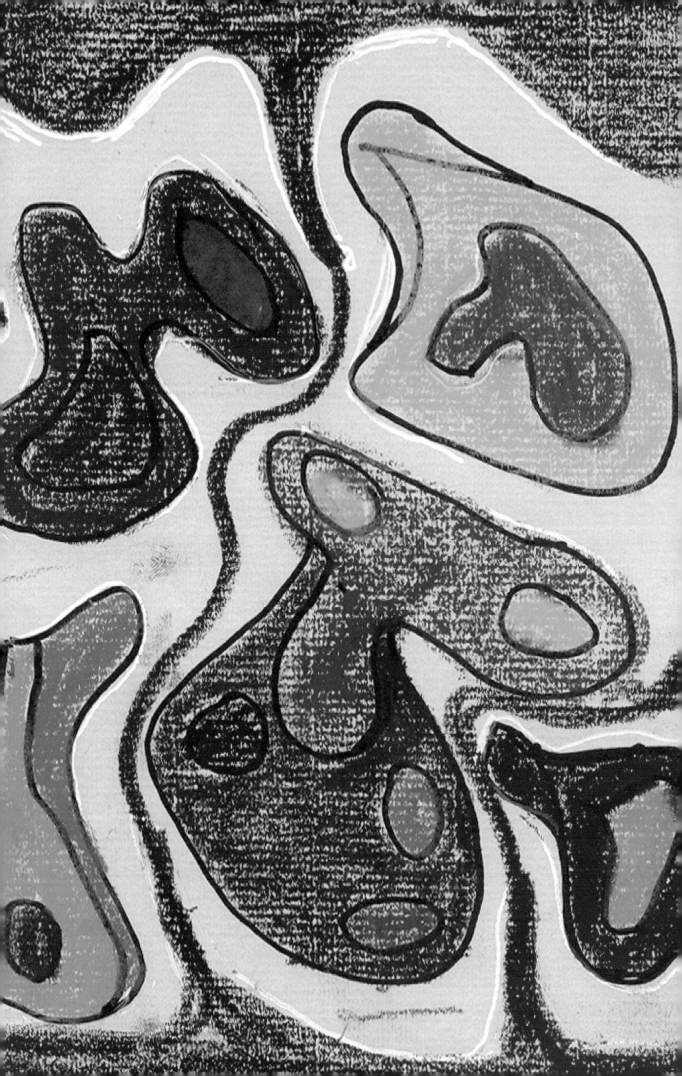

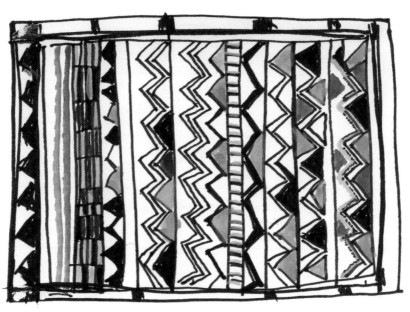

Requalification and Expansion of Thermal Complex

Montecatini Terme, Pistoia, Italy 2007– ongoing

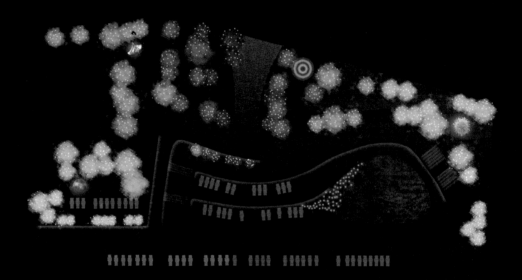

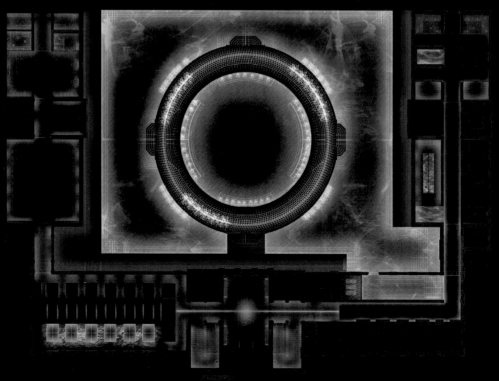

Intervention for the requalification and planning of the heritage building "Le Leopoldine" of the *Terme di Montecatini*.

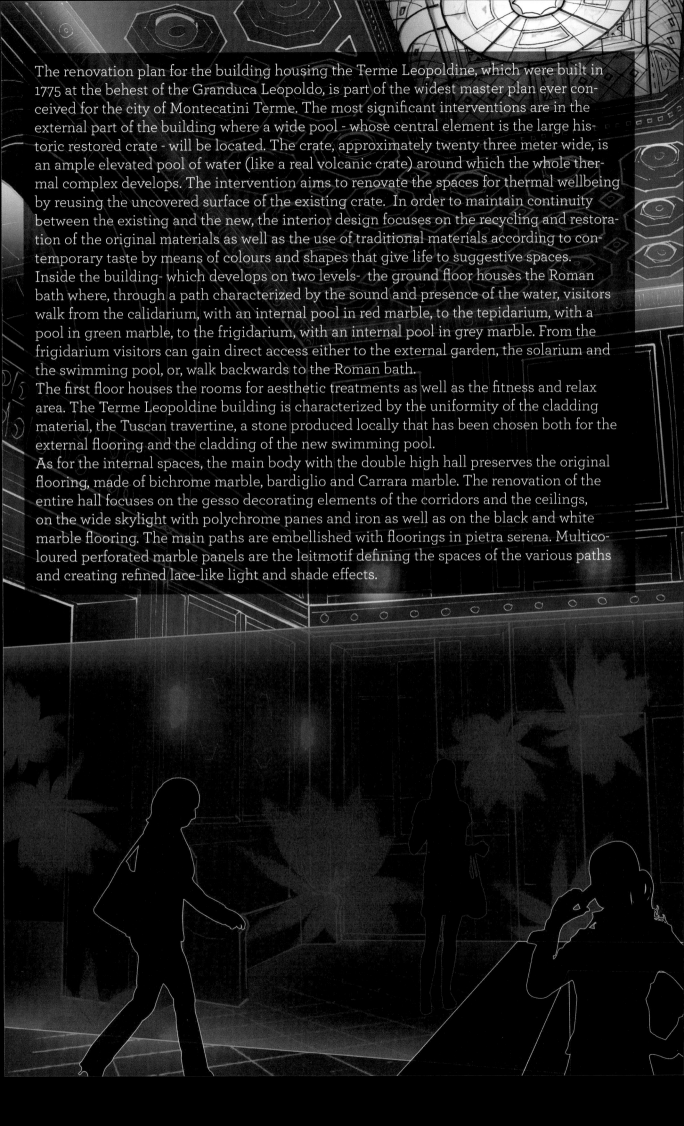

The renovation plan for the building housing the Terme Leopoldine, which were built in 1775 at the behest of the Granduca Leopoldo, is part of the widest master plan ever conceived for the city of Montecatini Terme. The most significant interventions are in the external part of the building where a wide pool - whose central element is the large historic restored crate - will be located. The crate, approximately twenty three meter wide, is an ample elevated pool of water (like a real volcanic crate) around which the whole thermal complex develops. The intervention aims to renovate the spaces for thermal wellbeing by reusing the uncovered surface of the existing crate. In order to maintain continuity between the existing and the new, the interior design focuses on the recycling and restoration of the original materials as well as the use of traditional materials according to contemporary taste by means of colours and shapes that give life to suggestive spaces.
Inside the building- which develops on two levels- the ground floor houses the Roman bath where, through a path characterized by the sound and presence of the water, visitors walk from the calidarium, with an internal pool in red marble, to the tepidarium, with a pool in green marble, to the frigidarium, with an internal pool in grey marble. From the frigidarium visitors can gain direct access either to the external garden, the solarium and the swimming pool, or, walk backwards to the Roman bath.
The first floor houses the rooms for aesthetic treatments as well as the fitness and relax area. The Terme Leopoldine building is characterized by the uniformity of the cladding material, the Tuscan travertine, a stone produced locally that has been chosen both for the external flooring and the cladding of the new swimming pool.
As for the internal spaces, the main body with the double high hall preserves the original flooring, made of bichrome marble, bardiglio and Carrara marble. The renovation of the entire hall focuses on the gesso decorating elements of the corridors and the ceilings, on the wide skylight with polychrome panes and iron as well as on the black and white marble flooring. The main paths are embellished with floorings in pietra serena. Multicoloured perforated marble panels are the leitmotif defining the spaces of the various paths and creating refined lace-like light and shade effects.

Business Garden Warszawa Hotel
Warsaw, Poland 2008

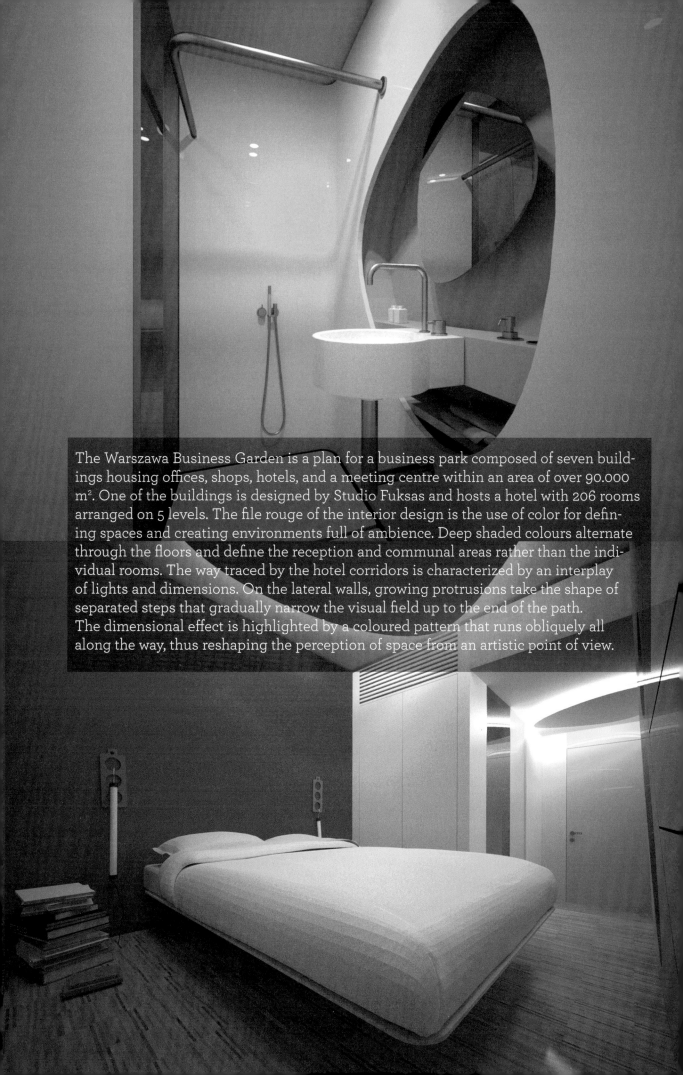

The Warszawa Business Garden is a plan for a business park composed of seven buildings housing offices, shops, hotels, and a meeting centre within an area of over 90.000 m². One of the buildings is designed by Studio Fuksas and hosts a hotel with 206 rooms arranged on 5 levels. The file rouge of the interior design is the use of color for defining spaces and creating environments full of ambience. Deep shaded colours alternate through the floors and define the reception and communal areas rather than the individual rooms. The way traced by the hotel corridors is characterized by an interplay of lights and dimensions. On the lateral walls, growing protrusions take the shape of separated steps that gradually narrow the visual field up to the end of the path. The dimensional effect is highlighted by a coloured pattern that runs obliquely all along the way, thus reshaping the perception of space from an artistic point of view.

Shenzhe

International Airport
Terminal 3_Shenzhen_China_2013

n Bao'an

The concept of the plan for Terminal 3 of Shenzen Bao'an international airport evokes the image of a manta ray, a fish that breathes and changes its own shape, undergoes variations, turns into a bird to celebrate the emotion and fantasy of a flight. The structure of T3 - an approximately 1,5-km-long tunnel – seems to be modeled by the wind and is reminiscent of the image of an organic-shaped sculpture. The profile of the roofing is characterized by variations in height alluding to the natural landscape.

The symbolic element of the plan is the internal and external double "skin" honeycomb motif that wraps up the structure. Through its double-layering, the "skin" allows natural light in, thus creating light effects within the internal spaces. The cladding is made of alveolus-shaped metal and glass panels of different size that can be partially opened. The passengers accede to the terminal from the entrance situated under the large T3 "tail". The wide terminal bay is characterized by white conical supporting columns rising up to touch the roofing like the inside of a cathedral.

On the ground floor, the terminal square allows access to the luggage, departures and arrivals areas as well as coffee houses and restaurants, offices and business facilities. The departures room houses the check-in desks, the airlines info-points and several help-desks.

The double and triple height spaces of the departures room establish a visual connection between the internal levels and create a passage for natural light. After checking in, the national and international passengers' flows spread out vertically for departures. The concourse is the airport key-area and is made up of three levels. Each level is dedicated to three independent functions: departures, arrivals and services. Its tubular shape chases the idea of motion. The "cross" is the intersection point where the 3 levels of the concourse are vertically connected to create full-height voids which allow natural light to filter from the highest level down to the waiting room set in the node on level 0.

The honeycomb motif is transferred and replicated on the interior design. Shop boxes, facing one another, reproduce the alveolus design on a larger scale and recur in different articulations along the concourse.

The interiors designed by Fuksas – placed in the internet-point, check-in, security-check, gates and passport-check areas – have a sober profile and a stainless steel finish that reflects and multiplies the honeycomb motif of the internal "skin". Sculpture–shaped objects - big stylised white trees - have been designed for air conditioning all along the terminal and the concourse, replicating the planning of amorphous forms inspired by nature.

This is also the case for the baggage-claim and info-point "islands".

Of Lithuanian descent, **Massimiliano Fuksas** was born in Rome in 1944. He graduated in Architecture from the University of Rome "La Sapienza" in 1969. Since the eighties he has been one of the main protagonists of the contemporary architectural scene.
He has been Visiting Professor at a number of universities such as: Columbia University in New York, the École Spéciale d'Architecture in Paris, the Akademie der Bildenden Kunste in Wien, the Staadtliche Akademia des Bildenden Kunste in Stuttgart. From 1998 to 2000 he directed the "VII Mostra Internazionale di Architettura di Venezia": "Less Aesthetics, More Ethics". Since 2000 he has been the author of the architecture column - founded by Bruno Zevi - in the Italian news magazine "L'Espresso". He is the recipient of several prizes and awards, including the "Légion d'Honneur" given by the President of the French Republic in 2010 and the "Commandeur de l'Ordre des Arts et des Lettres de la République Française" in 2000. He won the "Grand Prix National d'Architecture Française" in 1999 and the career prize "Vitruvio International a la Trayectoria" in 1998, Buenos Aires, Argentina. He is a member of RIBA - Royal Institute of British Architects, London, UK -, AIA - American Institute of Architects, Washington D.C., USA -, the Académie d'Architecture, Paris, France, and the Accademia di San Luca, Italy.

Doriana Fuksas was born in Rome where she graduated in History of Modern and Contemporary Architecture at the University of Rome "La Sapienza" in 1979. She has also earned a degree in Architecture from ESA, École Spéciale d'Architecture, Paris. She has worked with Massimiliano Fuksas since 1985 and has been director in charge of "Fuksas Design" since 1997. She has received a number of prizes and international awards. In 2013 she has been appointed "Commandeur de l'Ordre des Arts et des Lettres de la République Française" and in 2002 "Officier de l'Ordre des Arts et des Lettres de la République Française".

Studio Fuksas, led by Massimiliano and Doriana Fuksas, is an international architectural practice with offices in Rome, Paris, Shenzhen. With built projects across Europe, Asia and North America, Studio Fuksas is characterized by an innovative approach as well as interdisciplinary skills and experiences consolidated over three decades through the design of: masterplans, offices, residential buildings, infrastructures, cultural centres, leisure centres, retail developments, hotels, shopping malls, public buildings, interior design and product design.

Studio Fuksas main projects include:
National Archives of France in Pierrefitte sur Seine-Saint Denis, Paris, France, 2013; **Shenzhen Bao'an International Airport**, Terminal 3, Shenzhen, China, 2013; **Baricentrale**, competition won for the redevelopment of railway areas, Bari, Italy, 2013; **Moscow Polytechnic Museum,** Moscow, Russia, competition won in 2013; **New Rome Congress and Hotel Centre**, Rome, Italy, 1998 - 2014; **Tower for the New Directional Centre for Regione Piemonte,** Turin, Italia, 2001 - 2014; **Guosen Securities Tower**, 2010 - 2016; **Rhike Park**, Tbilisi, Georgia, 2010 – ongoing; **Chengdu Tianfu Cultural and Performance Centre,** Chengdu, China, competition won in 2012; **Tbilisi Public Service Hall**, Tbilisi, Georgia, 2012; **Hotel-Business Management School Georges-Frêche**, Montpellier, France, 2012; **Peres Peace House**, Jaffa, Tel Aviv, Israel, 2009; **San Paolo Parish Complex**, Foligno, Italy, 2009; **MyZeil shopping mall,** Frankfurt, Germany, 2009; **Armani Fifth Avenue**, NYC, USA, 2009; **Zenith Music Hall**, Strasburg, France, 2008; **New Milan Trade Fair,** Rho-Pero, Milan, Italy, 2005; **Ferrari Headquarters and Research Centre**, Maranello, Modena, Italy, 2004.